CALIFORNIA STATE UNIVERSITY

Earth Surface Systems

Series Editors
Andrew Goudie and Heather Viles

This series will provide accessible and up-to-date accounts of the physical and natural environment in the past and in the present, and of the processes that operate upon it. The authors are leading scholars and researchers in their fields.

Published

Caves
Processes, Development and Management
David Gillieson

The Changing Earth
Rates of Geomorphological Processes
Andrew Goudie

Land Degradation
Laurence A. Lewis and Douglas L. Johnson

Oceanic Islands
Patrick D. Nunn

Earth Surface Systems
Complexity, Order, and Scale
Jonathan D. Phillips

Humid Tropical Environments
*Alison J. Reading, Russell D. Thompson
and Andrew C. Millington*

Forthcoming

Rock Slopes
Robert Allison

Drainage Basin
Form, Process and Management
K. J. Gregory and D. E. Walling

Plant and Animal Introductions
P. J. Jarvis

Deep Sea Geomorphology
Peter Lonsdale

Holocene River Environments
Mark Macklin

Wetland Ecosystems
Edward Maltby

Arctic and Alpine Geomorphology
Lewis A. Owen, David J. Evans and Jim Hansom

Weathering
W. B. Whalley, B. J. Smith and J. P. McGreevy

Earth Surface Systems

Complexity, Order, and Scale

Jonathan D. Phillips

First published 1999

2 4 6 8 10 9 7 5 3 1

Blackwell Publishers Inc.
350 Main Street
Malden, Massachusetts 02148
USA

Blackwell Publishers Ltd
108 Cowley Road
Oxford OX4 1JF
UK

Library of Congress Cataloging-in-Publication Data

Phillips, Jonathan D.
   Earth surface systems: complexity, order, and scale / Jonathan D. Phillips.
     p. cm. – (The natural environment)
   Includes bibliographical references and index.
   ISBN 1-55786-934-0 (alk. paper)
   1. Environmental sciences. 2. Earth sciences. 3. System analysis. I. Title II. Series.
GE40.P48 1999
550–dc21   98-21952
                                                CIP

British Library Cataloguing in Publication Data

A CIP catalogue record for this book is available from the British Library.

Typeset in 10 on $11\frac{1}{2}$ pt Sabon
by Best-set Typesetter Ltd, Hong Kong
Printed in Great Britain by TJ International, Padstow, Cornwall

This book is printed on acid-free paper

To Lynn, for leading us to where the most important dreams always come true; to Bay, for being so gloriously independent of her initial conditions; and to Nate, for teaching me how to walk by faith.

Contents

Preface

To explain what this book is, or at least is intended to be, it is perhaps easier to begin by explaining what it is not. First, it is not a text purporting to expound the general principles of what earth surface systems are and how they function. The structure, processes, mechanics, and evolution of such systems have been dealt with by other, abler scientists; I particularly commend to you the books by Huggett (1985, 1991, 1995), and the classic by Chorley and Kennedy (1971) is still worth a look. For a primer on earth surface systems, any good physical geography text should suffice.

Second, this book is not a methodological treatise. There is a methodological chapter, and quite a bit of methodological material, but I present only enough of this to explain what I have done, to make (I hope) my major points and whence they were derived clear, and to allow the fascinated – or skeptical – reader to apply the methods to his or her own problems or to check my analyses.

Third, while much of the language (jargon?) and techniques of nonlinear dynamical systems theory is used herein, I do not attempt to place the study of earth surface systems within the confines of mainstream nonlinear science, at least not in the sense of applying methods used for abstract mathematical and laboratory systems to the wonderfully dirty and noisy real earth. The reason for this, expanded upon within the book, is that it is inappropriate to import, without modification, concepts and terminology from mathematics and physics to the far more complex systems of the real world. For example, most traditional definitions of deterministic chaos hinge on sensitive dependence upon initial conditions. Of what relevance is this to the *empirical* earth and environmental sciences, where initial conditions are not merely unknown, but manifestly unknowable? I believe a nonlinear dynamical systems approach to geomorphology, hydrology, pedology, and related sciences must be based soundly on concepts from those sciences. Therefore, traditional concepts and terminology of nonlinear science are in some cases adapted and modified to suit my (our) needs, and in other cases discarded in favor of (perhaps analogous) concepts and terminology mined directly from the geosciences.

Finally, I have not attempted to provide comprehensive coverage of the bewitching and bewildering variety of earth surface systems. I have attempted to provide variety, both in terms of biogeophysical phenomena and geographical coverage. But the topical mix is clearly and unavoidably biased by my training as a process-oriented soil, fluvial, and coastal geomorphologist, and by my field experience, which is heavily concentrated on the south-eastern coastal plain of the US. Having said what the book is not, let me turn to what it *is* . . .

From the time of my first exposure to general systems theory in Robert Giles's general systems ecology course at Virginia Tech nearly two decades ago, my scientific and scholarly interests have been systems-oriented. I have always been more interested in the whole than in the parts; whatever meager contributions I might have made to reductionist science were inspired by my perception that they were critical to a holistic understanding of some earth surface system. First, then, this book is a testament to my interest in the system-level behavior of earth surface systems. Many geoscientists adopt a systems framework in that they are careful to consider environmental contexts and interrelationships in efforts to comprehend a given process–response relationship or to unravel the history of some piece of the planet. Rather than (or in addition to) recognizing earth surface systems as a necessary tool for understanding the parts thereof, I am explicitly interested in the behavior of these systems as a whole.

Second, this book is an attempt to address a fundamental quandary – with philosophical, practical, and methodological dimensions – in the earth and environmental sciences. That is the simultaneous presence of order and complexity. Order, regularity, stability, and predictability exist at the same time and in the same places as disorder, irregularity, instability, and unpredictability. The emphasis on one or the other can be a function of purpose, epistemology, or spatiotemporal scale. The ubiquity of this situation led me to believe that there are fundamental, common if not ubiquitous, properties of earth surface systems which lead to this situation. I believe that I have shown this to be the case, and this book is an effort to convince the reader likewise.

Third, despite the variety and complexity of earth surface systems, there are some general principles about the way in which they function, and about the ways in which the interactions of the biosphere, lithosphere, hydrosphere, and atmosphere manifest themselves at the surface of our planet. This book purports to discover and elucidate some of those principles. Finally, the "holy grail" of my research efforts has been the development of a general theory of earth surface systems. This book is intended to be a step in that direction. I hope to be persuasive enough to convince at least a few others that such a thing is both possible and worth doing.

The audience I seek is geographers, geologists, hydrologists, soil scientists, and other earth and environmental scientists. Basic, introductory-level training in physical geography or interdisciplinary environmental science is all that is required to make use of this book, if I have written it well enough. Advanced training in pedology or geomorphology or geophysics or Quaternary science or whatever will enable the critical reader to dissect those portions of the book relevant to their disciplinary perspective and field experience. The mathematics may be unfamiliar to many, but the math is not complex. One advantage of the methodological approach I have chosen is that, despite its basis in nonlinear dynamical systems theory, it is closely linked to the far more familiar exercises of linear stability theory. Even more generally, the entire approach is based on analyses of systems depicted, directly or indirectly, as box-and-arrow models of interactions. The latter are, in at least a general sense, familiar to nearly all.

My acknowledgments must include some formal academic intellectual debts beyond those revealed in the citations. The aforementioned Dr Giles started me, for better or worse, down this path. Dr Chuck Ziehr, one of the finest teachers in the business, somehow imbued a former journalist with the confidence to tackle quantitative geography. Drs Karl Nordstrom and Bill Renwick overcame their skepticism of the (shall we say) unconventional views of a PhD student and supported my first substantive research efforts in earth surface systems. Richard Huggett provided encouragement to my first, abortive overtures about producing a book of this nature several years ago.

On a more personal level, my wife Lynn has suffered, more or less gladly, the foolishness of living with "Doctor Mud," complete with an unpredictable schedule and field boots which are never allowed past the garage. My son Nate has been a cheerful and helpful field companion and assistant; he has helped dig many a soil pit and bagged up many a sample. More importantly, his imagination and marvel are contagious. Finally, there is Bay Rochelle Phillips, born during the writing of this book, too early (by more than 15 weeks), and too small (at 753 grams, many of my sediment samples weigh more). Bay reminded me that complicated, inexplicable things may be part of clearer patterns that we just haven't seen yet, and that seeking the connections between the impossibly complex and the abundantly clear is as important a quest as there is.

Earth Surface Systems

The Nature of Earth Surface Systems . . .

Everything is connected to everything else. The world is indeed, as the bard asserted, a stage. But this stage is no mere platform upon which the drama of life, evolution, and change is played out. This is a stage which both influences, and is influenced by, the things which happen upon it. Moreover, in many cases the stage *is* the drama. Everything that happens in, on, to, or near the surface of the earth is connected, directly or indirectly, to everything else. The earth surface environment is an active and complex place, at the interface of the lithosphere, atmosphere, hydrosphere, and biosphere. It is marked by webs of interrelations, mutual adjustments, chain reactions, flows and cycles of energy and matter, feedbacks, and complex responses. To have any hope of understanding this stage, we have to consider those interactions, mutual adjustments, and feedbacks. But to make the problem manageable, we have to carve the vastness of the planetary surface environment into chunks we can handle. I call those chunks *earth surface systems*.

What is an earth surface system? Let's break it down. By "earth surface," I am referring to the surface *per se*, as well as the areas above and below which directly and proximately affect the surface environment. The earth surface environment in this context therefore includes the atmosphere, particularly the troposphere, and more particularly still the lower troposphere. Downward, the definition includes the earth's crust, although in the book our attention will usually be restricted to the outer portion of the crust above the weathering front (for example, the regolith).

By "system," I mean a set of interconnected parts which function together as a complex whole. The parts are the elements or components of the earth surface environment, which may include objects (for example, landforms, soil horizons, vegetation), stores of energy or matter (for example, soil moisture, organic matter, elevation), or processes (for example, infiltration, evapotranspiration, hydrolysis). The interconnections involve flows, cycles, and transformations of energy and matter. An earth surface system is thus a set of interconnected components of the earth surface environment which function together as a complex whole.

The emphasis will be on geomorphic, hydrologic, and pedologic systems. This is not to say that other types of earth surface system (ESS) are not important. The fact that there is some sort of emphasis arises from the unfeasibility of covering everything; the emphases I have chosen arise from my training and interests. There are systems treated in this book which are purely geomorphic (such as topographic evolution). Others are purely hydrologic (surface runoff generation) or pedologic (soil development). For the most part, however, the ESS include elements of at least two of the environmental spheres, and many of the systems truly integrate landforms, soils, hydrology, and biota, as well as other influences such as climate and tectonics.

The formal application of specific systems frameworks, such as nonlinear dynamical systems (NDS) theory or general systems theory, is relatively rare in the earth and environmental sciences. I call this the study of earth surface systems, where the primary goal is the holistic understanding of system-level behavior. However, a systems-oriented viewpoint or approach is quite common. I call the latter a systems framework, where systems concepts are employed to facilitate or to put in context reductionist studies of process–response relationships. The study of ESS and the systems framework typically (if, perhaps in the latter case, implicitly) have in common a concern with the following (Phillips 1992d):

- Emphasis on dynamic behaviors rather than description or classification.
- Concern with interactions in addition to the behavior of individual elements.
- Attention to mutual adjustments.

The study of ESS presupposes this type of systems viewpoint, but is *explicitly* rather than implicitly concerned with dynamics, interactions, and mutual adjustments; understanding these is the primary rationale for study. In the systems framework, dynamics, interactions, and mutual adjustments are usually of interest strictly to the extent necessary to understand a particular aspect of the system. The study of ESS essentially reverses the priorities of the systems framework, so that the system itself is the object of study, and the reductionist study of features within it is of interest only to the extent that they facilitate an understanding of system-level behavior.

Pahl-Wostl (1995) gives a good account in the context of ecology of how a nonlinear dynamical systems-oriented viewpoint similar to my own approach differs from more traditional reductionist approaches. Kellert (1993) provides extensive discussion of the different types of questions posed and answered by the NDS approach as compared to other approaches. Let me try to summarize it this way, with the understanding that these ideas will be explored more fully and rigorously later in the book. The complexities encountered in nature are typically part of a hierarchy. These complicated, irregular, sometimes random-appearing and inexplicable patterns are part of broader-scale regular, orderly patterns. The complexities also encompass smaller-scale, more orderly and understandable components. Thus the chaotic complexities of turbulent streamflows are part of a broad-scale order whereby the rate and direction of mean or net flow is quite predictable, and are composed of a great many individual particle trajectories which are individually well- understood and described by fundamental physical laws. The regular, orderly features found in nature are also typically part of a hierarchy. Simple, orderly patterns often arise from complex underlying dynamics, and are often, in the aggregate, part of broader-scale complex patterns. Thus, the simple geometry of beach cusps apparently arises from the complex nonlinear interactions between beaches and waves or from the complicated formation of edge waves. At broader scales, the cusps are only one part of quite irregular coastline geometries. Beneath the order lies chaos. And surrounding the chaos is order.

None of this is particularly new. Traditionally, however, scientists have concentrated on disassembling observed complexities into simpler, more orderly pieces. The ESS approach concentrates on the other axis: determining the broader-scale order arising from the complex dynamics. Likewise, traditional science has concentrated on using the regularities and orderly patterns observed in nature as building blocks to explain higher-level complexities and irregularities. The ESS approach again moves in the other direction, seeking to determine the nature of the complex dynamics underlying the orderly patterns.

... and the Earth Surface Systems of Nature

The sociology and politics of science, and the intellectual division of labor, often lead us to compartmentalize nature by disciplinary categories. Thus ecologists, who by definition should be concerned with the interactions between organisms and their environment, sometimes ignore the environment to concentrate on the interactions between organisms themselves. Likewise, geologists are likely to focus strictly on the lithosphere despite the obvious influences of (say) climate and vegetation on geological processes. Even physical geographers and pedologists, who by their paradigms and intellectual traditions should be well positioned to synthesize, tend to retreat to their subspecialties as fluvial geomorphologists or clay mineralogists or microclimatologists. We can, and have, learned a lot by this isolationist approach. For certain problems, and at appropriate spatial and temporal scales, it is acceptable – even necessary – to isolate a stream channel or a weathering profile from some of the factors which influence it, treating the latter as external controls or boundary conditions.

The next step beyond reductionism and isolationism is to consider two-way interactions: for example, the effects of climate on landforms, or of vegetation on soil erosion. These studies are also quite useful and have a rich tradition. Any investigation of pairwise interactions, such as the study of soil–climate relationships or the effects of topography on hydrologic response, can take our understanding of ESS only so far, however. There are two main reasons for this. First, the pairwise *relationships are often bidirectional*, not unidirectional. It is legitimate to study the effects of vegetation on soil erosion, of course, but ultimately the effects of soil erosion influence vegetation. Likewise, topography and channel network topology are important controls over and predictors of hydrologic response, but they are also responses to hydrologic processes. Second, the pairwise *interactions are often mediated by other environmental components*. Vegetation directly influences erosion by increasing surface resistance and dissipation of water or wind energy. However, indirect effects via infiltration, litter cover, and soil organic matter may be of equal or greater importance. Similarly, the impacts of erosion on vegetation may be direct, via physical damage, but are more likely to be indirect, mediated by the effects of erosion on soil properties, which then influence vegetation. For a final example, the exposure of high-albedo soil horizons may directly influence climate via changes in solar energy reflection and absorption, but changes in soil moisture storage and fluxes (and perhaps in vegetation, too) and runoff response may be of comparable or greater importance.

To stay with the soil erosion example, an understanding of a system characterized by erosion requires, first, an appreciation of the basic mechanics and process–response relationships involved. Second, it requires knowledge of the influence of individual factors, such as vegetation, hydroclimatology, soil properties, topography, and so on. Third, one must examine the mutual adjustments and interactions between erosion and the various controls and influences. Finally, one must ultimately come to grips with the fact that each major component of the system – climate, soils, vegetation, topography, hydrology, and so on – affects, and is affected by, the other components. At each or any stage, one may discover that understanding at some previous stage is inadequate. For example, in my own work on semi-arid land degradation systems (Phillips 1993b), I discovered that a basic understanding of the relationship between soil moisture and vegetation cover in dryland environments was (at that time) missing from the literature. Thus the emphasis on holistic, whole-system approaches here is not intended to relegate reductionist studies to a subordinate status. Such studies are not only a necessary precursor to the ESS approach, but provide an ongoing means to improve, refine, and even test systems-oriented models, explanations, and predictions.

The Music of the -Spheres

Earth surface systems invariably reflect the interactions between the four major spheres of the planetary environment. The *atmosphere* comprises the envelope of gases that surrounds the planet, mixes with elements of the other planetary spheres, and occupies the spaces or voids in soils, sediments, and rocks. The impacts of the atmosphere on other spheres, and vice versa, are typically manifested in terms of climate.

The *lithosphere* refers to the solid, inorganic material of the earth, principally rock, soil, and sediment. From the ESS perspective, the most important aspects of lithospheric processes are those manifested at the surface, and the predominance of impacts of the other spheres on the lithosphere occur at or near the surface. Further, the interface of the lithosphere with the other spheres is restricted to the crust. Therefore the interactions between the litho- and other spheres are largely expressed in terms of landforms and the weathered outer "skin" of the earth.

The *hydrosphere* (including the frozen *cryosphere*) encompasses water in all its forms. The flows and cycles of water, as well as its changes between solid, liquid, and gaseous states, comprise the single most important set of mechanisms for interactions between the spheres. The list of intersphere interactions that do not directly or indirectly involve H_2O would be a short one.

The *biosphere* is composed of earth's living things and organic matter. The influence of the inorganic spheres on the biosphere is profound, as any farmer or herdsman could attest. The impacts of the biosphere on the inorganic spheres can be equally profound, up to and including the exertion of important controls on the very composition of the atmosphere, and effects on the lithosphere to the point that in some cases much of the regolith can be referred to as a biomantle.

Example 1: Fluvial geomorphology

Any number of examples can be cited to illustrate the importance of the interactions between the atmo-, litho-, hydro-, and biospheres. Here I choose three. First, consider figure 1.1, from a textbook on fluvial geomorphology. Here A. D. Knighton (1984) depicts the major interrelationships in the fluvial system. Even the most restricted and reductionist considerations of fluvial geomorphology must incorporate elements of the hydrosphere and lithosphere. In figure 1.1, the atmosphere (climate) and lithosphere (geology, basin physiography) are clearly shown as critical independent, external controls on drainage basins. The biosphere is another important

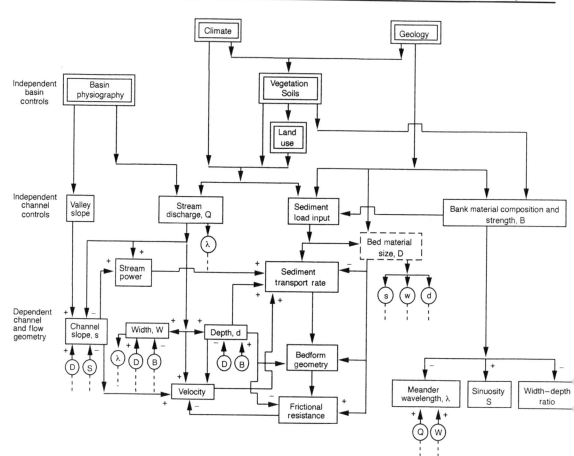

Figure 1.1
The fluvial system as depicted in Knighton (1984).

external control (vegetation, soils, land use), which is itself partly controlled by climate and geology. Numerous elements of the hydrosphere (stream discharge and flow characteristics) and lithosphere (form and material properties of channels) are shown.

It is important to recognize three additional points about figure 1.1. Like most ESS, fluvial systems are characterized by numerous *mutual adjustments*, where two or more components both affect and are affected by each other: for instance, channel width and meander wavelength. Also, figure 1.1 represents a simplified *abstraction*. A large (infinite?) number of other aspects of the environmental spheres could be incorporated into a model or study of fluvial systems. For example, biosphere influences could also be represented via the effects of riparian vegetation on bank strength and composition. Finally, the notion of independent versus dependent factors rests on *spatial and temporal scale*. At millennial time scales fluvial processes in stream channels partially determine

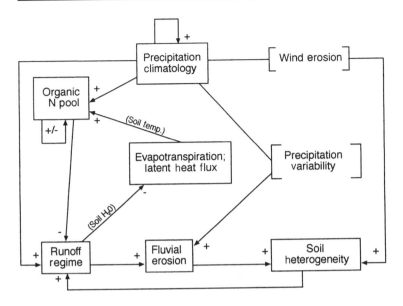

Figure 1.2
Interactions between major system components in semi-arid land degradation or desertification. Components in boxes are the critical system components; those in brackets represent intermediate linkages. Adapted by Phillips (1993b) from Schlesinger et al. (1990).

basin physiography, rather than the latter serving as an external control. And at broad spatial scales the hydrologic transfers of rivers are important influences on climate, rather than simply a response.

Example 2: Semi-arid land degradation

Second, consider figure 1.2 showing the critical interactions involved in semi-arid land degradation (desertification) processes. The basic model of interactions was devised by Schlesinger et al. (1990), and adapted by Phillips (1993b) for mathematical analysis. For a long time most studies of dryland degradation phenomena focused on two-component cause-and-effect relationships, such as the effects of overgrazing on vegetation cover or soil erosion, or the influence of baring high-albedo soils on precipitation. A large number of such studies produced an overwhelming wealth of information, but with sometimes conflicting implications, and in dire need of synthesis. Schlesinger and others (1990) strove to isolate and aggregate the most critical components and the feedback interactions between them. The atmosphere (precipitation climatology, wind, latent heat flux), lithosphere (fluvial erosion, soils), hydrosphere (evapotranspiration, precipitation, runoff), and biosphere (organic nitrogen, soils) are all explicitly represented.

In figure 1.2 it is noteworthy that several system components incorporate more than one sphere. The evapotranspiration/latent heat flux component, for example, incorporates the atmo-, hydro-, and biospheres (the latter via transpiration). It is also apparent that the components are very broadly defined, in that a number of specific variables might be used to describe them. Soil heterogeneity, for example, might be expressed in terms of spatially distributed variables describing soil physical, chemical, mineralogical, and biological properties, soil taxonomy, indices of soil development, and many others. Finally, the analysis of the system depicted in figure 1.2 shows that it is inherently unstable: perturbations are likely to persist or grow, and small changes may lead to disproportionately large impacts. This trait is common in ESS, and will be revisited in due time.

Example 3: Global hydrologic cycle

Finally, consider (figure 1.3) a conceptual model which, at first glance, is restricted to the hydrosphere. This model of the global hydrologic cycle, however, inescapably includes the atmo-, litho-, and biospheres. Two reservoirs and five fluxes directly involve the atmosphere, which is clearly critical in the hydrologic cycle. The lithosphere exerts important influences over the surface and underground water reservoirs, and the flux to the oceans; it in effect comprises the "buckets" which hold these reservoirs. The biomass reservoir and transpiration processes explicitly include the biosphere.

Many of the points made about figures 1.1 and 1.2 apply to the global hydrologic cycle. The additional point to be made is that even at the broadest scales and highest levels of aggregation the

Figure 1.3
The major reservoirs and fluxes in the global hydrologic system (from Committee on Opportunities in the Hydrologic Sciences 1991).

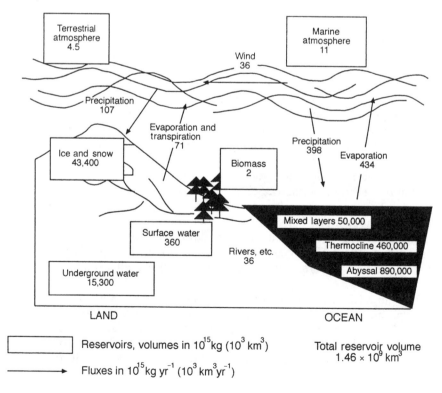

Reservoirs, volumes in 10^{15} kg (10^3 km^3)

Fluxes in 10^{15} kg yr^{-1} (10^3 km^3 yr^{-1})

Total reservoir volume
1.46×10^9 km^3

complex interactions of the atmo-, litho-, hydro-, and biospheres are evident and inescapable. It is these interactions which produce the planetary environments we occupy, use, enjoy, and study – the earth surface systems. That is why I think of ESS as "the music of the -spheres. "

Of Spades and Basins

The controls over process–response relationships vary with spatial and temporal scale. The distinction between dependent and independent variables is a function of scale. The tools we use, the questions we ask, and the answers we get all vary with spatial and temporal scale. Scale linkage – efforts to synthesize phenomena which operate over a broad range of spatiotemporal scales – is a pervasive, critical problem in the environmental sciences. Further, I will argue later in this book that some fundamental qualitative properties of ESS, such as stability and instability, are a function of scale. It seems, therefore, that a word or two is in order about spatial and temporal scales as they appear in this book.

We will touch upon all spatial and temporal scales from soil micromorphology and grain-to-grain interactions on sand dunes and in stream channels, up to global-scale plate tectonics and evolution. In general, however, I will deal with systems which range in spatial scale from approximately the size of a spadeful of soil to that of a large drainage basin (and some of the latter, of course, can be very large). The temporal scales I treat range from instantaneous up to those of Quaternary geology (roughly two million years). One reason for limiting my scales is that I can't cover (and am not conversant with) everything, so that is one way of limiting my scope. Another arises from my personal conviction that this spatial-scale range falls within that which is most relevant to humans. We experience, use, and manage ESS as plots, fields, hillslopes, channels, basins, and ecosystems. Information at smaller scales – cellular, molecular, atomic, subatomic – is certainly useful and often indispensable to the understanding of ESS. But we cannot hope to gain anything approaching a complete understanding of ESS based on what we learn at those scales. Likewise, the workings of ESS (no doubt) fit into some cosmic framework. But we cannot comprehend earth surface systems through science conducted at astronomic and cosmic scales. To understand earth surface systems we require knowledge at scales ranging from those of the geochemist, soil physicist or microbiologist to those of the tectonophysicist or global climate modeler. I have staked out my scales of interest – from spades to basins – within this range.

The Last of How It Is

This book is somewhat nontraditional in several ways. Traditionally, science seeks quantitative predictions and is believed to reduce uncertainty. The work I report here, despite the mathematical analysis, makes qualitative predictions and discovers uncertainties. Traditional science focuses on processes and mechanics, while I focus more on structure, organization, and patterns of flux. Such holistic, nonlinear dynamical systems (NDS) approaches to science are becoming increasingly common – skeptics might even say trendy. But because the skeptics are many, and the NDS-based approach still relatively new, it is worthwhile to compare and contrast my approach with that of traditional reductionist science. Further, because much of what you'll find here differs quite a bit from "mainstream" nonlinear science, we need to identify our niche in that regard as well.

Let's use an analogy of an earth surface system as a mechanical machine (many philosophers, most postmodernists, and some nonlinear scientists really hate these mechanical analogies, but that's part of the fun of employing them). And let's introduce three scientists: the reductionist, the systematist, and the nonlinearist. The traditional, *reductionist* approach to understanding this machine would involve taking it apart, examining each component, and determining the form and function of each part (or even grinding it up or dissolving it and analyzing the chemical elements, but we won't take it that far). Once an agreeably complete understanding of the parts seems within view, the traditional reductionist scientist would begin to look at the linkages between individual parts: how and why this wire connects to that lead, or this chain goes on that sprocket. Eventually, if the machine is simple enough and has no unknown parts, if each part works in only one way and each component-to-component relationship works only one way, if the whole machine behaves and functions according to the sum of its parts, and if the environment in which the system functions or the rules governing its operation remain reasonably constant, a complete and thorough understanding of the machine will be available (phew). We can predict its performance, control its operation, and fix it if it breaks. If there are unidentified parts, if the components work in several ways, if part-to-part relationships are variable, if the whole is more than (or different from) the sum of the parts, or if the environment or rules change, our understanding will be quite incomplete. We may not be able to predict or control the machine, or to fix it if it is broken.

The systems approach would involve carefully observing the intact machine. The *systematist* would catalog all the components and how they are linked together. She would observe the machine in operation, and systematically (pun intended) measure the flows and cycles of energy and matter into and out of the machine, and among the system components. She would observe or experiment to determine how the machine functions under different environmental constraints or rules of operation. Eventually, if done well, this approach will yield a complete and thorough understanding of the machinery if, that is, one can identify all the important components and the relationships among them, and measure the energy/matter fluxes, without taking the machine apart. The systematist can predict its performance, control its operation, and if it is broken, will know what part needs attention (although without the help of the reductionist she may not know enough to repair the part or to build a new one).

The *nonlinearist* would start with a set of functional equations describing the machine's operation (probably generated by reductionists) or a network model (probably produced by systematists) or maybe a humongous data set produced by either. He (notice the politically correct alternating pronoun) would then determine whether, or under what circumstances, the machine produces stable output, cyclical output, or wildly varying, apparently random output. If the other scientists have been unable to come up with a complete part list or network diagram or understanding of the machine's functions, the nonlinearist can provide important clues and guidance for achieving those goals. He can also perhaps devise ways to predict the machine's functional state or output in the absence of complete and thorough knowledge of the machine, though the predictions may be probabilistic. Unfortunately, the nonlinearist will only be able to give us vague information about the machine's components. If it is broken, he can probably tell us whether or not it can be fixed, but not how to go about the repairs. If we know little or nothing about the machine he may be able to tell us things about its behavior that traditional scientists could not, and even how many components it probably has. However, the nonlinearist will not be able to tell us what those components are.

The reductionist, systematist, and nonlinearist above are crude caricatures, of course. But, as professional wrestling teaches us, crude caricatures have their place. Here they illustrate not just different approaches, but the different kinds of knowledge generated. These forms of knowledge are complementary, not competitive. It should be clear by now that I believe the reductionist approach can only take us so far in understanding earth surface systems. But reductionist knowledge often provides the underpinnings, starting points, data, or measurement technologies for nonlinear and systems approaches (reductionist knowledge is also often useful in its own right, but that's in somebody else's book).

I also believe the nonlinearist approach can only take us so far in understanding complex earth surface systems. Yes, it can be used to gain a toehold of understanding for complex ESS that defy traditional reductionist analysis. Certainly, NDS analysis can generate new and different types of understanding. But understanding system trajectories in phase space (for example) really brings us no closer to understanding the way ESS work than does reductionist knowledge of the system components and their processes. The analysis of complex nonlinear systems must be, I believe, firmly linked to what's happening on and in the ground and water and air.

Behold, then, the systematist. She needs input from the reductionists to fully understand the components of her system models and the links between them. She needs the nonlinearist to understand the dynamics of the complex systems she's piecing together. Likewise, the reductionist must depend on the systematist to place his work in perspective; to put the pieces of the puzzle together. The nonlinearist leans on the systematist to provide some grounding, to link system states and trajectories and phase spaces and such to real things happening in real places.

I presume here to join the ranks of the systematists. I gratefully accept the painstaking work of my reductionist colleagues as my building blocks and puzzle pieces. I eagerly adopt and adapt the analytical tools and holistic perspectives of my nonlinearist colleagues as my means of getting a handle on how these complex systems work. And now the preliminaries are out of the way – let's rock.

2

Qualitative Analysis of Earth Surface Systems

This is the chapter where we get into the ugly business of methods, models, and mathematics. Among nonscientists (and a surprisingly large number of scientists) there is rampant "math anxiety" and antipathy toward equations. Even among scholars of a statistical or mathematical bent, there is a wide lack of interest in concepts and models of little direct relevance to one's own favorite family of equations and modeling tools. This is, perhaps, a defense mechanism against the awesome and impossible task of understanding all equations one can be bombarded with. I have, therefore, tried to strike a balance. I give what I hope is enough information (and certainly there is enough if the references cited are consulted) to allow the keenly interested or the keenly skeptical to dissect and/or reproduce the analyses in the book. On the other hand, I have provided *only* what is necessary to accomplish the above. Finally, I have made a concession to those who are uninterested in the equations or who are willing to assume (for the sake of argument, of course) that I know what I am doing in this regard. Thus I will include, in language as plain as I can manage, simple statements of what the equations mean and what they are good for. Thus, equation-haters can still make some use of the rest of this book.

Qualitative Analysis of Partially Specified Dynamical Systems

Dynamical systems are dynamic in that they are not static: they change and evolve. Further, attaching the *-al* means that these systems can exhibit changes in system state; that is, in the basic, characteristic status or situation of the system. Simple examples of state changes include river channels alternating between degrading and aggrading states; glaciers alternating between advancing, re-treating, and stationary states; and salt marshes changing between states where the surface elevation change is greater than, less than, or approximately equal to the rate of sea-level rise. ESS are clearly dynamic, and the examples above – along with innumerable others one might envisage – shows that they are dynamical, as well.

Partially specified systems means that there is not complete information about the web of interactions. In a biogeochemical mass balance system, for instance, we may know the mass in each of the major reservoirs at a given time, but not the magnitude of the fluxes between them. A more common situation, and one we will focus on, is where the direction or basic nature of system interactions are known, but the specific magnitudes (or the specific governing functional equations) are unknown. For example, we may know that soil moisture has a negative link to wind erosion (higher soil moisture leads to less wind erosion, and vice versa), but we cannot attach a number or an equation to that link.

We deal with partially specified systems in the study of ESS for two main reasons. First, in many cases that is all the information we have. It is relatively easy to completely specify a given relationship over a definite time in a particular place. It is exceedingly difficult to completely specify more than a few links in multiple locations and at different times. The measured, quantified, completely specified relationship between soil moisture and wind erosion rates that we obtained on the plains of Oklahoma during a storm in March 1998 will not be transferable to the steppes of Russia or the deserts of Australia or the sand dunes of Ireland or the Patagonian pampas, or perhaps even to our Oklahoma site during a storm in November. However, we can be confident that at any of those places, at any time, greater soil moisture will be associated with less susceptibility to wind erosion, and drier soil will blow away more readily. Naturally, these generalizations are all-other-things-being-equal, and all other things almost never are. However, in this framework they (the other things) *are* explicitly accounted for elsewhere in the model, and need not hinder our partial specification of the soil moisture/wind erosion link. The second, related, reason for dealing with partially specified systems is for generalization. On those rare occasions when one has a fully specified complex system, those specifications, in detail, have only a limited spatial and temporal range of application. However, as generalizations – whether components have positive, negative, or negligible influences on each other – those links are widely applicable.

The "qualitative" comes in partly because the systems are partially specified. We cannot get quantitative results, but we can determine whether systems are stable or unstable, potentially chaotic or nonchaotic, self-organizing or not, and many other things. We can identify the states they may be in, and under what conditions they may be in them. Qualitative questions are the most important, anyway. It would be nice to know precisely how much forage a rangeland will have, and exactly what the runoff and erosion rates from it will be. But the truly crucial question is whether the range condition will improve, decline, or remain constant under various management plans or environmental changes.

Stability and Lyapunov Exponents

Now that I have, rather lengthily, explained what is meant by qualitative analysis of partially specified dynamical systems, let's get to it. The starting point is to assume we've got an earth surface system with n components $x_{i,j} = 1, 2, 3, \ldots n$. Each component x_i at least potentially affects, and is affected by, each of the other x_i. Thus,

$$dx_i/dt = f_i(x_1, x_2, x_3, \ldots, x_n)(c_1, c_2, c_3, \ldots, c_n) \qquad (2.1)$$

where c_i represents the functional relationships (which in some cases may be zero).

A small perturbation is denoted by δx, and in our context is defined as one which is not sufficient to alter the basic feedback relationships in the system (i.e. to change the "rules"). For instance, in a stream channel cross-section an increase in flow where the water remains within the channel would be a small perturbation. During overbank flow the "rules" are different; such a flow increase would constitute a large perturbation. Note that if a system is stable with respect to small perturbations it may be unstable with respect to large ones. A system that is unstable to small perturbations is unstable to any perturbation.

Growth rates of δx are given by a set of linear differential equations of the form:

$$d\delta x_i/dt = A_{ij}\delta x_i \qquad (2.2)$$

The A_{ij} are elements of the Jacobian matrix of $f = (f_1, f_2, f_3, \ldots, f_n)$. They are given by

$$A_{ij} = \partial f_i(x_1, x_2, x_3, \ldots, x_n)/\partial x_j|_{x_0} \qquad (2.3)$$

where x_0 is the state of the system before the perturbation. Mathematically, x_0 is an equilibrium state, but equilibrium can be a loaded term, particularly in geomorphology (Thorn and Welford 1994). In our context x_0 is any persistent, nontransient predisturbance state.

The analysis of small perturbations around a (mathematical) equilibrium in this manner is, in effect, a linearization via a Taylor series analysis (Puccia and Levins 1985: 244–8). This procedure is standard in linear stability analysis. It is also useful in the analysis of *nonlinear* dynamical systems because the stability properties of the linearized system are exactly the same as for the original, nonlinear parent system.

The Jacobian matrix can be conceptualized as an interaction matrix. This matrix has complex eigenvalues, the real parts of which are the Lyapunov exponents (λ). A dynamical system has n Lyapunov exponents. The λ give the rate of system convergence or divergence over time following the perturbation:

$$\mathbf{x}(t) = \mathbf{v}\mathbf{C}e^{\lambda t} \tag{2.4}$$

where $\mathbf{v}$ are the eigenvectors and $\mathbf{C}$ the initial conditions. All quantities above except t are vectors. If the eigenvalues/Lyapunov exponents are negative, the deviation is exponentially damped, and the pre-perturbation state is restored. If $\lambda > 1$, the perturbation grows exponentially over some finite time. If any eigenvalue/ Lyapunov exponent is positive, the system is unstable. If all $\lambda < 0$, the system is stable. If any $\lambda = 0$ and all others are negative, the system is neutrally stable.

How can we determine this for a partially specified system? Think of the interaction matrix $\mathbf{A}$, with entries a_{ij}. Each entry is simply $+$, $-$, or 0, representing the positive, negative, or negligible influence of the ith component on the jth component. Using the Routh–Hurwitz criterion, this is all the information required to determine whether there are any postive λ.

The eigenvalues of $\mathbf{A}$ are the roots of the characteristic equation

$$a_0\lambda^n + a_1\lambda^{n-1} + a_2\lambda^{n-2} + \ldots + a_{n-1}\lambda + a_n = 0 \tag{2.5}$$

The as are coefficients. The characteristic equation can be rewritten in terms of feedback (as demonstrated by Puccia and Levins 1985, 1991). Feedback at level k of a system (F_k) represents the mutual influences of system components on each other for all loops with k components. For $k = 2$, for example, the feedbacks represent bivariate relationships of the form $a_{ij}a_{ji}$, $i \neq j$, or $a_{mm}a_{kk}$, $m \neq k$; $m,k \neq i,j$. Feedback at $k = 3$ includes all loops involving three components, and so forth. Only disjunct loops, sequences of one or more a_{ij} with no common component i or j are included.

$$F_k = \Sigma(-1)^{m+1}Z(m, k) \tag{2.6}$$

where $Z(m,k)$ is the product of m disjunct loops with k components. By convention, $F_0 = -1$. Then the characteristic equation becomes

$$F_0\lambda^n + F_1\lambda^{n-1} + F_2\lambda^{n-2} + \ldots + F_{n-1}\lambda + F_n = 0 \tag{2.7}$$

The Routh–Hurwitz criteria give the necessary and sufficient conditions for all real parts of all eigenvalues to be negative (and thus for all Lyapunov exponents to be negative). These are:

- $F_i < 0$, for all i.
- Successive Hurwitz determinants are positive.

Only alternative determinates have to be tested, and the second condition can be expressed algebraically as follows:

$$F_1F_2 + F_3 > 0, \text{ for } n = 3 \text{ or } n = 4 \tag{2.8a}$$

$$F_4F_1^2 + F_1F_5 - F_1F_2F_3 - F_3^2 > 0, \text{ for } n = 5 \text{ or } n = 6 \tag{2.8b}$$

For $n > 6$ the mathematics become complicated and unwieldy. Fortunately, there are ways of eliminating or aggregating system components without affecting stability properties, and other means of assessing some dynamical properties for systems of seven or more components. If the Routh–Hurwitz criteria are not met, all eigenvalues and Lyapunov exponents are not negative, and the system is not stable. Book-length treatments of these methods are given by Puccia and Levins (1985) and by Logofet (1993). Explanations and applications in geomorphology and hydrology are given by Slingerland (1981) and Phillips (1987a, 1990d).

In summary, if you can translate or depict an earth surface system as a box-and-arrow diagram or interaction matrix specifying the components and whether they have positive, negative, or negligible impacts on each other, the stability of the system can be determined using the Routh–Hurwitz criteria (RHC). It does not matter if the system is nonlinear – it probably is – because the stability properties of the original nonlinear system and the interaction-matrix linearized version of it are identical. The RHC allows one to determine whether or not the system has any positive Lyapunov exponents. If it does, that indicates that the system is unstable to small perturbations and potentially chaotic (see below).

Instability and Chaos

The n-dimensional space occupied by a dynamical system is its state space; an m-dimensional $(m < n)$ space embedded within the state space is called the phase space. We usually speak of phase rather than state spaces because we are often unsure whether we have truly identified every component of the system. At any time, the system state is given by some location within the phase space; linking these points defines a trajectory. Attractors are points or areas within the phase space which control the trajectories, and to which all trajectories are drawn. There are many types of attractors, both simple and complex. The most complex are called strange or chaotic attractors.

Deterministic chaos is a property sometimes exhibited by non-linear dynamical systems whereby even simple deterministic systems can produce complex, pseudorandom patterns in the absence of (or independently of) stochastic forcings or environmental heterogeneity. In chaotic systems, irregularity, complexity, and unpredictability are an inherent trait of system dynamics. Chaotic systems exhibit sensitivity to initial conditions in that initially similar states diverge exponentially, on average, and become increasingly different over time.

Mathematically, chaotic systems are often said to have sensitive *dependence* on initial conditions because even the most minuscule differences in starting values (attributable to nothing more than computer rounding errors at many decimal places) lead to divergent results later. This terminology has been unfortunately misleading for earth scientists, as it superficially implies the ability to estimate or infer initial conditions from the present state of a system. What a "magic bullet" that would be for paleoenvironmental reconstructions! From the ESS perspective, however, the sensitivity to initial conditions is better understood as *independence* of initial conditions. This is because even minor, geologically and ecologically insignificant variations in starting conditions could lead to a very different system state later on!

Chaotic systems are also sensitive to perturbations of all magnitudes. Sensitivity to initial conditions is usually stressed in the nonlinear dynamical systems (NDS) and chaos theory literature because of the predominance of numerical modeling. In the study of ESS, as opposed to numerical models, initial conditions are not just unknown, but unknowable. And because we deal with long time periods, numerous perturbations are likely to have occurred. Sensitivity to perturbations is therefore at least as important as sensitivity to initial conditions from our vantage point. If a soil landscape is characterized by chaotic pedogenesis, for example, that means the evolution of the soil is not only sensitive to minor variations in, say, parent material texture or mineralogy, but to any of numerous small perturbations along the way (an anthill or gopher hole, a patchy fire, a tree-fall and so on).

Sensitivity to initial conditions is depicted by this standard relationship from chaos theory, where the Δs represent the difference between two system states at the start (time zero) and at some future time t:

$$D_t \sim \Delta_0 e^{\lambda t} \tag{2.9}$$

Look familiar? Note that the separation at time t is a function of the Lyapunov exponent, just as the growth of perturbations is in equation 2.4. The system is not, and cannot be, chaotic unless there is at least one positive Lyapunov exponent. As an unstable system has at least one $\lambda > 0$, dynamic instability is tantamount to a chaotic system. In much of the NDS literature, a chaotic system is in fact defined as one which has a positive Lyapunov exponent. The link between qualitative stability and deterministic chaos is explored in some detail with respect to geomorphic and pedologic systems by Phillips (1992e, 1993d).

A useful identity is

$$\Delta \lambda_i = \Sigma a_{ii} \tag{2.10}$$

The a_{ii} represent self-effects within the system. Negative self-effects, for example, are shown by soil moisture storage, which is limited at saturation points by finite capacity and by increased flow, and at low levels by the wilting point. An example of positive self-effects would be a situation where vegetation exhibits negative density dependence (such as seed source effects) such that low densities stay low due to a lack of seed stock, and increasing densities encourage further increases. Mathematically, nonzero self-effects occur where

$$\delta(dx_i/dt)/\partial x_i|_{x_0=a_{ii}} \tag{2.11}$$

This will come into play later, in the discussion of self-organization.

In summary, deterministically chaotic systems are sensitive to minor variation in initial conditions and to small perturbations, such that minuscule changes or variations grow over time. The presence of chaos is indicated by a positive Lyapunov exponent: therefore the possibility of chaos may be assessed using the qualitative stability analysis described earlier.

The Largest Lyapunov Exponent

Lyapunov exponents are a staple of NDS and chaos theory. They not only indicate the presence or absence of chaos, but also measure the rates of convergence or divergence in NDS. For the moment, consider a nonlinear dynamical ESS as a small n-dimensional sphere of initial conditions. As time progresses, the sphere evolves into an ellipsoid whose principal axes contract or expand at rates given by the spectrum of Lyapunov exponents λ_i. The Lyapunov exponents may be arranged in order from the most rapidly expanding to the most rapidly contracting, such that

$$\lambda_1 \geq \lambda_2 \geq \ldots \geq \lambda n$$

When the attractor is chaotic, the system trajectories in phase space diverge, on average, at an exponential rate given by the largest exponent (λ_1) (Wolf et al. 1985).

If one has a completely specified system, the entire Lyapunov spectrum can be calculated numerically (Wolf et al. 1985). In ESS, however, one is working with field evidence and this approach is not applicable. A convenient property of any NDS is that the dynamics of the entire system can be deduced from a single observable, such as topography in the case of a geomorphic system, or discharge in the case of a watershed.

Rosenstein et al. (1993) showed that randomly chosen pairs of initial conditions diverge exponentially (on average) in a chaotic system at a rate given by the largest Lyapunov exponent:

$$d(t) = Ce^{\lambda_1 t} \tag{2.12}$$

Note the similarity to the equations describing growth of small perturbations and convergence or divergence of initial conditions over time. Here $d(t)$ is the average divergence of randomly selected pairs at time t and C is a constant that normalizes the initial separation. Rewriting this way

$$\lambda_1 = \ln d(t) - \ln C \tag{2.13}$$

illustrates the dependence of the largest Lyapunov exponent on the average rate of divergence over time.

Envisage a landscape with an initial set of elevation differences between discrete points or areas, given by C. If topographic relief increases over time, the average divergence is positive, $\lambda_1 > 0$, and the system is chaotic. If relief declines, the elevation differences become smaller on average, and the system is nonchaotic, with a negative λ_1 and thus all exponents negative (this example is treated in some detail by Phillips, 1995b). On a shorter time scale, consider an experimental plot where cells have some initial variation in infiltration rates. By measuring the extent to which these rates diverge or converge, on average, by time (t), you could measure $d(t)$ and thus estimate λ_1.

Therefore, the dynamics of an NDS can be inferred from a single observable (elevation, stream discharge, infiltration rate). The rate of divergence (if any) of randomly selected pairs is given by the largest Lyapunov exponent, and the largest Lyapunov exponent can be estimated by measuring the average convergence or divergence of randomly selected pairs.

Spatial Chaos

Temporal chaos in the presence of anything other than perfect isotropy – and the real world is never perfectly isotropic – will result in spatial chaos. The latter is manifested as increasing spatial complexity over time. Imagine a landscape with tiny variations in initial conditions and/or characterized by small perturbations here and there. If evolution is chaotic, those small changes and variations will get larger, and the spatial variability must increase. Symbolically,

$$\Delta_{p,q}(t) \sim \Delta_{p,q}(0)e^{\lambda t} \tag{2.14}$$

where $\Delta_{p,q}$ represents the difference between any two locations p,q at time zero and after time t. The basic dynamical system equation can also be rewritten in spatial form, as

$$dx_i/dy = f_i(x_1, x_2, x_3, \ldots, x_n)(b_1, b_2, b_3, \ldots, b_n) \tag{2.15}$$

Here, dy represents change in space rather than change in time, and the bs are coefficients. Phillips (1993f) discusses spatial chaos in some detail, and shows that traditional spatial statistics are unable to distinguish deterministic chaos from random or stochastic noise. The bottom line is that *temporal chaos begets spatial chaos. If an ESS is characterized by chaotic evolution, spatial complexity will increase over time.*

Entropy and Information

The "chaoticity" of a dynamical system can be measured by the Kolmogorov (K-) entropy, discussed in book-length detail by Kapitaniak (1988), and in the context of soils and landforms by Culling (1988b). In an idealized, abstract world where every system is either deterministic and nonchaotic, chaotic, or random (white noise), the K-entropy indicates the state of the system. In the first, K-entropy is zero; in the last, it is infinite. Positive, finite K-entropy indicates deterministic chaos. In practice, of course, positive finite entropy can be produced by various combinations of orderly determinism, white noise, and chaos. K-entropy is linked to Lyapunov exponents in that the sum of the positive λ equals the K-entropy.

In the ESS literature one is likely to encounter not only the K-entropy of NDS theory, but physical and statistical entropies. Physical entropy is associated with the thermodynamic state of a system. Statistical entropy measures or reflects the degree of unpredictability of a system, and negative entropy is equated with information or reduction in uncertainty. Entropy calculations are identical in all three cases, though the mathematical and thermodynamic entropies have opposite signs. The statistical entropy of an observed realization of an ESS is equivalent to the K-entropy of the underlying dynamical system (Oono 1978; Culling 1988b; Zdenovic and Scheidegger 1989; Fiorentino et al. 1993).

It is possible to decompose the observed entropy of a spatial pattern in the landscape into those components associated with the combination of information and noise, and with deterministic chaos. Let's start with a spatial pattern of cells, each with a discrete occurrence of a spatially distributed feature B, which might be soil or rock type, land surface unit or vegetation type. The entropy is given by

$$H = 1/N \ln \left(N! \bigg/ \prod_{1=1}^{n} f_j! \right) \tag{2.16}$$

where N is the number of cells and f_j the number of cells associated with the jth outcome of B.

Now introduce feature A, a spatially distributed variable that exerts some control or potential influence on the outcome of B. Thus A is a control factor, such as parent material or elevation, which influences the response factor B, such as soil texture or microclimate. We then have a distribution of N locations or cells, each with one of $i = 1, 2, \ldots, n$ outcomes of A and $j = 1, 2, \ldots,$

m outcomes of *B*. Things are simpler if one assumes (as I once did: Phillips 1987b) that *A* is static, with a fixed distribution. Things are closer to reality if the control as well as the response variable is allowed or assumed to occur in different spatial patterns. Then *H(A)* and *H(B)* are the maximum entropy for the spatial distributions of *A* and *B*, which would occur if any value or outcome could exist in any location or cell:

$$H(A) = 1 / \left[N \ln \left(N! \Big/ \prod_{i=1}^{n} f_i! \right) \right] \qquad (2.17a)$$

$$H(B) = 1 / \left[N \ln \left(N! \Big/ \prod_{j=1}^{m} f_j! \right) \right] \qquad (2.17b)$$

Phipps (1981) devised a means for measuring the entropy of the distribution of response factor *B*, as constrained by the control factor *A*:

$$H_A(B) = 1 / \left[N \ln \left\{ \prod_{1=1}^{n} \left(f_i! \Big/ \prod_{j=1}^{m} f_{ij}! \right) \right\} \right] \qquad (2.18)$$

If *A* and *B* were completely independent, a pure noise signal would result, with $H_A(B) = H(B)$. If the pattern is completely deterministic, such that there is only one possible outcome of *B* for a given *A*, then $H_A(B) = 0$. The possible presence of deterministic chaos is thus indicated by

$$0 < H_A(B) < H(B).$$

In practice, the factorials in the entropy equations become unmanageable with large *N* and *f*. In that case, the equations below produce reliable esimates.

$$H(A) = 1 / [N \ln(N \ln N) / \Sigma(f_i \ln f_i)] \qquad (2.19a)$$

$$H(B) = 1 / [N \ln(N \ln N) / \Sigma(f_j \ln f_j)] \qquad (2.19b)$$

$$H_A(B) = (1/N) \ln[\Sigma f_j \ln f_j - \Sigma f_i \ln f_i] \qquad (2.20)$$

If $H_A(B) < H(B)$, the uncertainty and entropy have been reduced by a factor (*r*) associated with the constraint *A* exerts on *B*:

$$r = \left(e^{NH(B)} \right) / \left(e^{NH_A(B)} \right) \quad \text{or} \quad \ln r = NH(B) - NH_A(B) \qquad (2.21)$$

The ln *r* is called mutual information in information theory, and measures the quantity of information transmitted without noise. Significance can be tested using the chi-square statistic, since ln *r* ~ χ^2. Entropy reduction is given by ln *r*/*N*. The use of mutual information functions in symbolic patterns or sequences of nonlinear dynamical systems is discussed by Li (1990).

If $0 < H_A(B) < H(B)$, it could indicate chaos, colored noise, or both. We can partition the entropy into its chaos and colored noise components by starting with the maximum possible entropy, that associated with a white noise pattern. We then subtract the order or information associated with A's constraints on B, and finally subtract $H_c(B)$, the entropy produced by chaotic dynamics:

$$H(AB) = [H(A) + H(B)] - [H(B) - H_A(B)] - H_c(B) \qquad (2.22)$$

Two endpoint cases can be identified. One is trivial: any outcome of A can occur anywhere, and all the information in the $H_A(B)$ channel is due to chaos, i.e. $H_c(B) = H_A(B)$. The other is where the control factor's pattern is fixed and/or completely determined (i.e. only one possible distribution). Then $H(A) = 0$, $H(AB) = H_A(B)$, and $H_c(B) = 0$. Between the extreme cases, A is neither fixed nor random, but is constrained or affected by some underlying factor a. Then the observed entropy of A is

$$H_{obs}(A) = 1/N \ln(xN!/\Pi f_i!), \qquad (2.23)$$

where $1/x$ is a factor by which the number of possible arrangements of A has been reduced, $1/x \geq 1$. There are many ways x might be calculated or estimated: one is by determining the mutual information of the joint distribution of A and a, in that case $1/x = r_{a,A}$. The maximum and observed entropies of A are related by

$$H_{obs}(A) = H(A) + \ln x. \qquad (2.24)$$

Because $H(AB) = H_A(B) + H_{obs}(A)$, we can substitute and simplify to obtain

$$H_c(B) = -\ln x \qquad (2.25)$$

The deterministic contribution to the entropy of the joint distribution of A,B is therefore a direct function of the extent to which A is constrained by some underlying control. For example, suppose that depositional environment constraints reduce the possible distribution of parent material textures (the control factor) by 87.9 percent ($x = 0.121$). Then the chaotic contribution to the entropy of the joint distribution of soil type and vegetation community (the response variable) is 2.112. This example comes from data on soil A-horizon and parent material textures from the North Carolina coastal plain (Phillips 1994). In that example deterministic "chaos" accounted for about 46 percent of the observed entropy of the pattern; the remainder is colored noise. This example will be fleshed out below.

I have used the example of one control and one response factor and one underlying constraint on the control factor. One can generalize to any level of hierarchy, by

$$H(k, k+1, k+2, \ldots, q) = \sum_{k=1}^{q} H(k) - \sum_{k=1}^{q} [H(k) - H_{k+1}(k)] - H_c(k_i)$$
$$(2.26)$$

$H_c(B)$ might well be called *deterministic uncertainty* rather than chaos. Classic deterministic chaos cannot be removed or resolved by simply gaining more or better information, while the underlying constraint x could, at least in theory, be so treated. In practice, however, there are many situations where some underlying effect is known, but cannot be measured. For example, individual trees are known to influence soil properties, yet in a soil thousands or more years old there is no way to ever know, other than for the present and the recent past, where and when a tree might have existed. It is feasible, however, to estimate the degree of underlying influence from tree effects (x), based on observations. What x and $H_c(B)$ have in common with classic chaos is that deterministic factors produce an increase rather than a decrease in uncertainty.

In summary, the entropy of a nonlinear dynamical system indicates its "chaoticity," because K-entropy is equal to the sum of the positive Lyapunov exponents. In real-world ESS, measured entropy can be due to deterministic complexity, or to "colored noise," the combination of randomness and deterministic order. In spatial distributions where one or more response factors are constrained or influenced by one or more control factors, it may be possible to decompose the deterministic complexity and colored noise components of entropy. The deterministic uncertainty or chaos is found to be a direct function of the extent to which the control factor is limited in its distribution by some underlying constraint. This spatial-domain deterministic uncertainty differs from classic chaos in some ways, but is similar in that completely deterministic factors or controls within an ESS increase, rather than decrease, uncertainty and unpredictability.

Self-organization

Self-organization is common in ESS. Flat or irregular beds of sand in stream channels or on aeolian dunes self-organize, under the influence of water and wind flows, into regularly spaced ripples of remarkably similar size and shape, for example. Patterned ground develops in periglacial environments, cellular water circulation and associated regularly spaced beach cusps form on shorelines, and fluvial erosion carves landscapes into orderly hierarchies of channels and basins. Self-organization is linked to instability and chaos, and can be described using the K-entropy. This argument is spelled out in greater detail in Phillips (1995a).

In a self-organizing system, K-entropy would have to decrease, at some level, to reflect an increase in order and organization. However, ESS include irreversible processes (such as photosynthesis and respiration, gravity-driven mass fluxes, and weathering), so

total entropy must increase over time (production of thermodynamic entropy by ESS processes results in an increase in statistical
and K-entropy of the landscape; see Brooks and Wiley 1988;
Culling 1988b; Zdenovic and Scheidegger 1989; Phillips 1995a).

If positive entropy is to increase over time in the absence of
stochastic forcings, a geomorphic system must have at least one
positive Lyapunov exponent because the K-entropy is equal to the
sum of all the positive Lyapunov exponents. However, if the negative entropy (information) at some level is to simultaneously increase, the sum of the negative λ must exceed the sum of the
positive λ; i.e. the average of the exponents must be negative. The
Lyapunov exponents are, you will no doubt recall, equal to the real
parts of the complex eigenvalues of the system. While the positive
Lyapunov exponents express the K-entropy or "chaoticity" of the
system, the negative λ reflect the exponential approach of initial
states to an attractor. Put another way, $\lambda > 0$ give the rate of
information loss, while $\lambda < 0$ give the rate of information gain.
If a system is to be self-organizing, there must be at least one
positive Lyapunov exponent, but the sum of λ must be negative.
Only then can the rate of information gain exceed the rate of
entropy production.

The argument can also be made without recourse to any links
between Kolmogorov and thermodynamic entropy. If an earth
surface system is undergoing spontaneous organization, initially
similar forms or features become more differentiated, on average,
over time. Thus a planar bed develops ripples or dunes, weathered
debris is differentiated into soil horizons, or a landscape is dissected
by fluvial erosion, to give a few examples. This differentiation
represents increasing divergence over time, on average. The latter
requires a positive Lyapunov exponent, and thus finite, positive K-
entropy. The increasing divergence over time cannot continue
indefinitely: witness the finite amplitude of bedforms, the matura-
tion of soil profiles, and erosion to base level. This means that
though the phase space stretches exponentially in one direction
according to the largest positive λ, the overall volume of the phase
space must contract. For the latter to be the case in a nonlinear
dynamical system, the sum of all n Lyapunov exponents must be
negative.

The sum of the diagonal elements of the interaction matrix is
equal to the sum of the real parts of the eigenvalues and thus to $\Sigma\lambda$.
The interaction matrix, even in qualitative form, can show not only
whether there are any positive exponents, but whether $\Sigma\lambda < 0$. The
diagonal elements of the matrix are the self-effects. Where only
negative self-effects (feedback) exist, or where they are stronger
than any positive effects, the sum of Lyapunov exponents is negative. Where positive self-effects dominate, $\Sigma\lambda > 0$, and the system is
strongly chaotic and not self-organizing. If there are no self-
reinforcing or self-damping feedbacks, or if they cancel each other
out, the sum of the Lyapunov exponents is zero and the system
is neutrally stable.

Thus, subject to two assumptions, we have two criteria for determining whether an ESS is self-organizing:

- The system is a (probably nonlinear) dynamical system of the form:

$$dx_i/dt = f_i(x_1, x_2, \ldots, x_n), (c_1, c_2, \ldots c_n)$$
$$i = 1, 2, \ldots, n \quad \text{and}$$

- The system can be represented by an $n \times n$ interaction matrix A.

Then, the two criteria for self-organization are:

- The matrix A is unstable according to the Routh–Hurwitz criteria.
- The sum of the diagonal of A is negative ($\Sigma a_{ii} < 0$).

In short, for an ESS to be self-organizing, entropy at some level must decrease over time to reflect an increase in order. But total entropy must increase because ESS involve irreversible processes. These can occur simultaneously only in a system which has a positive Lyapunov exponent, but where the sum of the Lyapunov exponents is negative. This can be tested using the methodologies described earlier. If a system is unstable by the Routh–Hurwitz criteria, and the sum of the diagonal elements is negative, it is self-organizing – otherwise, it is not.

Summary

While similar biophysical laws and principles govern (or at least influence) ESS, the variety of conditions on the planet dictate that fully quantitative specifications are location specific. The complexity of ESS, and their temporal variability, dicate that even location-specific specifications can be exceedingly difficult to obtain. We can circumvent these difficulties, and generalize beyond specific locations or systems by using partially specified systems. If we can identify the critical components of an earth surface system in the form of a box-and-arrow diagram or interaction matrix specifying the components positive, negative, or negligible influences on each other, we can determine:

- Whether the system is stable, unstable, or neutrally stable in response to small perturbations.
- If the system is potentially chaotic, and thus sensitive to initial conditions and small disturbances.
- If the ESS is self-organizing.

We can also use NDS methods, adapted for the situations encountered in the study of ESS, to examine empirical data. This enables us to:

- See whether initially similar points, on average, become more or less similar over time, and thus whether there is a positive Lyapunov exponent and deterministic chaos.
- Examine spatial patterns to estimate the amount of deterministic uncertainty.

We will employ, and expand upon (just a little!), these methods in ensuing chapters to learn a bit more about ESS and how they work. I give a couple of examples from my own previous work below, with specific calculations linked to the equations above.

Example 1: Wetland response to sea-level rise

The response of coastal features such as salt marshes to sea-level change is a good example of the problem geoscientists are frequently confronted with. The problem involves complex systems with numerous interrelationships among system components. While the general nature of the relationships between the important components is known, the exact form of the governing equations, and the quantitative specification of the interactions (beyond site- and time-specific data sets) is unknown. I have chosen this problem to illustrate some of the methods described precisely because it is typical of the kind of problem amenable to these methods, because my own interest in marsh response to sea-level rise goes back to my doctoral dissertation, and (not incidentally) because I have dealt with this example before (Phillips 1992e).

Coastal wetlands such as salt marshes are geologically ephemeral and sensitive to sea-level change. Marshes may grow or decline in response to a number of geomorphic, hydrologic, climatic, and human-induced stimuli directly, indirectly, or coincidentally associated with sea-level change. Along many of the world's coastlines relative sea level is rising, and the literature on marsh response to coastal submergence is synthesized by Orson et al. (1985), Stevenson et al. (1988), Nichols (1989), Reed (1990), and Kearney et al. (1994).

The implications of deterministic complexity in wetland response to sea level are important beyond attempts to understand and predict geomorphic and ecological changes. Coastal wetlands and their stratigraphy are often used as indicators of Holocene sea level and associated climate trends (Rampino and Sanders 1981; Kearney et al. 1994), and these wetlands are of critical interest to environmental managers because of their important ecological, economic, and aesthetic roles in coastal ecosystems.

Complex spatial patterns of marsh shoreline erosion and marsh interior deterioration have been mapped and measured in, among others, the Delaware Bay, Chesapeake Bay, and southern Louisiana estuarine systems (Phillips 1986c; Kearney et al. 1988; Nyman et al. 1993). To what extent are these complex, irregular patterns attributable to local variations in environmental controls and stochastic forcings versus inherent complex deterministic dynamics of the marsh system?

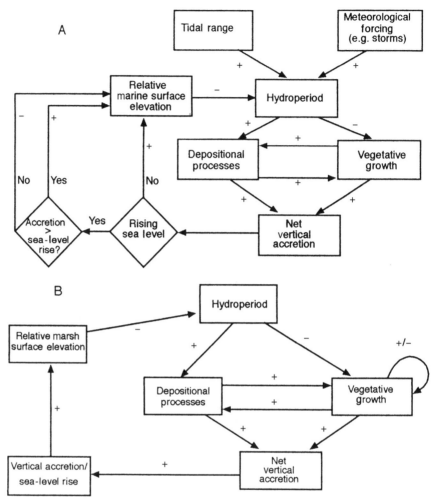

Figure 2.1

Interactions between sea-level rise, hydroperiod, and organic and inorganic deposition processes controlling response of coastal marshes to sea-level rise. (A) Redrawn from Reed (1990); (B) adapted from Reed (1990) in Phillips (1992e) by excluding external forcings, limiting consideration to rising sea level, and by adding vegetation self-effects.

Denise Reed (1990) has synthesized current knowledge of the important components involved in marsh response to sea-level rise and the relationships between them (figure 2.1). The information is qualitative in the sense that the direction of the relationships is known, but no dynamic equations or generally applicable numbers can be attached to them. Longer inundation regimes (hydroperiods), for instance, are known to stimulate sediment deposition on the surface and inhibit vegetative production, but these relationships cannot be specifically parameterized for a general model. The links in the model are described in detail by Reed (1990) and summarized by Phillips (1992e).

The stability properties of a system such as that shown in figure 2.1A are unaffected by removing components which function strictly as external forcings; that is, which have interaction arrows leading to other components, but which are not themselves influenced by other components. Thus tidal range and meteorological

Table 2.1 Interaction matrix for the marsh response system shown in figure 2.1

	RMSE	HP	Veg.	Dep.	NVA	VA/SL
RMSE	0	$-a_{12}$	0	0	0	0
HP	0	0	$-a_{23}$	a_{24}	0	0
Veg.	0	0	$\pm a_{33}$	a_{34}	a_{35}	0
Dep.	0	0	a_{43}	0	a_{45}	0
NVA	0	0	0	0	0	a_{56}
VA/SL	a_{61}	0	0	0	0	0

RMSE, relative marsh surface elevation; HP, hydroperiod; Veg., vegetative production; Dep., depositional processes; NVA, net vertical accretion; VA/SL, vertical accretion v. sea-level rise ratio.
Source: Phillips 1992e.

forcings can be removed from Reed's model for analytical purposes. By treating accretion versus sea-level rise as a ratio, only one arrow from this component to relative marsh surface elevation is necessary. The system can be further simplified by considering only the case of rising sea level. I have also added the self-effects due to density dependence known to influence vegetation. After these changes, figure 2.1B is obtained. This is translated into an interaction matrix in table 2.1. Note that all entries in the table are simply negative $(-a_{ij})$, positive (a_{ij}), or negligible (0).

The first Routh–Hurwitz criterion is that all $F_k < 0$. For the matrix in table 2.1 the application of equation 2.6 gives us

- $F_1 = a_{33}$ or $F_1 = -a_{33}$; $F_1 > 0$ or $F_1 < 0$, depending on whether vegetation self-effects are positive (for example, seed source effects at low densities) or negative (competition at high density).
- $F_2 = a_{34}a_{43} > 0$.
- $F_3 = 0$.
- $F_4 = 0$.
- $F_5 = -a_{12}a_{24}a_{45}a_{56}a_{61} < 0$.
- $F_6 = (-a_{12}a_{24}a_{43}a_{35}a_{56}a_{61}) - [(\pm a_{33})(-a_{12}a_{24}a_{45}a_{56}a_{61})]$. F_6 will be negative if $a_{33} < 0$, or if the following holds: $(-a_{12}a_{24}a_{43}a_{35}a_{56}a_{61}) > a_{33}F_5$.

The first Routh–Hurwitz criterion is thus violated by F_2, F_3, and F_4. This is sufficient to demonstrate instability, potential chaos, and a positive Lyapunov exponent. F_1 and F_6 could be positive or negative depending on the nature of vegetation self-effects. If the stability analysis hinged on this, field evidence would have to be relied on to determine stability in particular situations.

The second criterion ($F_4F_1^2 + F_1F_5 - F_1F_2F_3 - F_3^2 > 0$ for this case) becomes material only where the first is met, but is generally judged by evaluating the relative strengths (typically judged as rates or magnitudes) of the interactions involved. For a case where all $F_k < 0$, one can see that the second and third terms on the left side of the inequality would have to outweigh the first and fourth terms.

Sometimes this is simple to judge; sometimes it is (given available information) impossible. This approach is quite robust and widely applicable, but no analytical technique works everywhere and all the time!

The response of marshes to coastal submergence is unstable and chaotic. Minor variations in initial conditions (such as underlying pre-Holocene topography) or small perturbations (such as local vegetation destruction) would become increasingly magnified over time. Given the spatial manifestations of chaos, this implies increasing spatial complexities in marsh shoreline configurations, and in the spatial mosaic of high marsh, low marsh, unvegetated pans, and open water. Such changes have been documented in several situations (Boston 1983; Phillips 1989g; Nyman et al. 1993, 1994; Kearney et al. 1994).

Marsh self-organization

For self-organization to occur in ESS, there must be progressive differentiation between initially similar points, but this differentiation must occur within finite and well-defined limits. As discussed above, this can only occur when there is a positive Lyapunov exponent (divergence and differentiation) but the sum of the Lyapunov exponents is negative.

In the marsh response system the sum of the Lyapunov exponents depends on the sign of the vegetation self-effects. If a_{33} is positive, this indicates that vegetation increases or decreases are self-reinforcing, independently of any other effects. For increasing vegetation cover, this might occur where seed sources and rhizome extension accelerate vegetation expansion. For declining vegetation, such effects might be associated with limited reproduction due to diminished seed production or allelopathy as numbers decline. Such a system is not self-organizing. This behavior could presumably exist only for the time scale of vegetation trends.

If vegetation is self-limiting (for example, due to competition at high densities), $a_{33} < 0$, and the sum of the Lyapunov exponents is negative. In this case the system is self-organizing. The unstable divergence would be resolved into patterns at broader scales. This type of behavior has been documented in a number of marshes around the world, where unstable growth of salt pans or open water enlarges these features, creating broader-scale self-organized patterns of vegetated marsh, pans, and open water (Pethick 1974; Boston 1983; Phillips 1986c; Nyman et al. 1993, 1994).

Example 2: Soil texture

Surface soil texture is affected by a number of factors. Foremost, or at least first, is the texture of, or inherited from, the parent material. This may be modified or influenced by weathering, age, translocation, erosion, deposition, bioturbation, and additions of mass. A typical view is that there is a systematic relationship between parent

Table 2.2 Contingency table showing joint occurrences of A-horizon and C-horizon textural classes

A-horizon texture	C-horizon texture						
	S	FS	LS	LFS	SL	FSL	SCL
S	14	0	1	0	0	0	0
FS	4	4	0	1	0	0	0
LS	11	0	5	0	3	0	4
LFS	4	3	2	5	1	4	3
SL	2	0	2	0	1	0	3
FSL	5	0	8	5	4	4	13
L	3	0	2	2	1	2	12
SiL	0	0	1	0	0	4	4

S, sand; FS, fine sand; LS, loamy sand; LFS, loamy fine sand; SL, sandy loam; FSL, fine sandy loam; L, loam; SiL, silt loam; SCL, loam or finer, mostly sandy clay loam.
Source: Phillips 1994.

material and surface horizon texture, upon which is superimposed stochasticity associated with the many additional factors which can alter particle sizes and their relative abundance. This view is expanded by a nonlinear dynamical systems perspective to allow for deterministic uncertainty.

In previous work (Phillips 1994) I presented a data set of 147 sample pedons from eastern North Carolina. For each, the data set included the surface (A-horizon) textural class, and the textural class of the lowermost C-horizon. In the unconsolidated sediments of the North Carolina coastal plain, C-horizon textures correspond to those of the parent material. From this, table 2.2 was prepared. Each entry represents the number of joint occurrences of A- and C-horizon textural classes, or the f_{ij} of equations 2.16–2.19.

The theoretical maximum entropy of the joint distribution (calculated using equation 2.19) is 5.38. This is the value that would occur if both surface and C-horizon textures were completely random. This theoretical maximum is considerably reduced by the obvious control of parent material texture; the entropy of surface texture as constrained by parent texture is 0.0528. To what extent is the observed entropy due to stochastic complexity (many degrees of freedom), and to what extent to underlying but unmeasured controls over parent material texture (deterministic uncertainty)?

Texture of the parent material is at least partially constrained by the depositional environment of the parent coastal plain sediments. The latter are of four basic types within the study area: modern uplands underlain by mixed marine, coastal, and estuarine deposits; alluvial and lower, younger marine terraces underlain by mixed alluvial, coastal, and estuarine sediments; relict barrier island deposits of mainly sandy beach and dune sediments; and modern alluvial floodplains. Relationships between depositional environments and C-horizon texture in the study area suggest that the number of possible arrangements of C textural classes is reduced by a factor of 8.26 (or r in equation 2.21 is equal to 1/8.26 or 0.121).

Referring to equation 2.25, the chaotic component of entropy is 2.112. Compared to the mutual information of the joint pattern of parent material and surface horizon texture (equation 2.21), the ratio of information generated by the chaotic component to that associated with the textural relationship is 0.84. The observed entropy of the joint distribution of surface and parent texture is 0.771. Less than 7 percent of this is due to the entropy of the pattern of surface texture constrained by parent texture. The remainder is generated by the deterministic component.

The analysis shows that, for the study area, deviations from the orderly relationship between surface and parent texture are affected but little by the many degrees of freedom in modifying soil texture. Rather, the majority of the uncertainty is associated with underlying deterministic controls on parent texture itself! There is a good deal more to be said about deterministic uncertainty. Some of it has been said by McBratney (1992), some in the article containing the example analysis above (Phillips 1994) and subsequent work (Phillips et al. 1996), and some in chapter 4.

3

Order and Complexity

The Order of Things

Order and regularity is pervasive in earth surface systems. Other-wise, engineers could not use equations to design dams, canals, and bridges. Soil surveyors could not use soil-landscape relationships to map soils. Geographers could not produce plausible maps of land-forms, ecosystems, or climates. You get the idea. Certainly, models, equations, maps, classifications, and similar generalizations are all simplifications of earth surface reality. They all mask, ignore, or smooth over irregularities, deviations, aberrations, anomalies. But the very concept of aberrations and anomalies affirms the concept of some underlying norm or order.

P. A. Burrough (1983) and W. E. H. Culling (1987, 1988a), among others, discuss the simultaneous presence of both order and disorder or irregularity in the landscape, and how, according to one's purpose, resolution, or epistemology one or the other might be of paramount concern. But troubling questions arise. Why do order and complexity always seem to exist simultaneously? Which is more important? Are the irregularities to be interpreted simply as deviations from some orderly plan, or are they important and informative in their own right? Is order merely apparent, the result of lumping and aggregation, and/or human cognitive processes, and/or the presence of similar, repetitive patterns within the com-plexities? We'll confront these questions. But first, let us examine some samples of order (and in a bit, complexity) in ESS.

The hydrologic cycle and the wisdom of God

There is a fixed and finite amount of H_2O in the earth–atmosphere system, and 98 percent of it is in the oceans. On, in, and over land, it is in constant flux, changing in both state and location. Yet these fluxes are, in at least a broad sense, quite predictable. And no matter how the system is perturbed, and what kinds of outrageous droughts, floods, storms and other changes occur, the cycling of water just keeps on going. This constancy and order has long fascinated and impressed humans.

The hydrologic cycle was initially rejected by Christian theologians, during a time when new ideas about nature were scrutinized by the church. However, the theory of the hydrologic cycle grew in favor when it became clear that it could be used to support the doctrine of a divine plan (Tuan 1968). Theologians saw in the cycle verification of their doctrine that earth was created expressly as the home of humans, and that all of its processes were part of a great ordered plan. Tuan (1968) shows that the hydrologic cycle as presented by seventeenth-century natural theologians differs but little from simplified versions of the cycle depicted in twentieth-century textbooks. The structure and function – indeed, the very presence – of global cycles of water, carbon, nitrogen, sulfur, and other substances is perhaps the broadest and most intuitively convincing evidence that order is pervasive in ESS despite the tricky details of hydrologic processes or biogeochemical transformations.

The runoff hydrograph

Within a drainage basin, there are often dramatic spatial variations in the factors that control the runoff response. Soils, slopes, vegetation, land use, antecedent moisture, and other factors may, in fact, exhibit quite complex spatial patterns. Further, the dynamics of precipitation, infiltration, subsurface and surface flow can also be quite complicated. Yet, despite all this, there is a relatively simple and straightforward relationship between the amount of precipitation that falls within the catchment and the amount of water that flows out of it.

Hydrologists are able to construct instantaneous unit hydrographs (sometimes based on empirical rainfall and streamflow data, and sometimes constructed based on the basin's topology and topography) which can predict the discharge output per unit of precipitation input. They are able to estimate runoff tolerably well using simple empirical methods with just a few variables, such as the runoff curve number. Often, a little number crunching will yield a perfectly satisfying one- or two-parameter equation which does a pretty good job of indicating how much runoff will result from a given precipitation event (the "rational method," as it is called in hydrology textbooks).

Granted, none of those technologies is perfect. All have wide confidence intervals, are prone to break down under certain hydrologic circumstances, and tell us little about hydrologic processes. But the fact that they work as well as they do – well enough for many engineering, planning, and water-management applications – is testament to the fact that some simplicity, or order, exists in the relationship between precipitation and runoff. Whether the relatively simple relationship arises from simply "averaging out" the complexities within the basin, as a scale-related function of system structure, or due to an overwhelming influence of precipitation as a determinant of runoff within a given setting, is a debate for another time. For now, suffice it to say that an orderly relationship – and quite a useful one at that – clearly exists.

The geometry of the landscape

Sometimes the order of earth surface systems is apparent in geometric patterns on the landscape (plate 3.1). The unmistakable geometry of the barchan dune, for example, will be found wherever there is sufficient sand and wind, and insufficient vegetation. Se-

Plate 3.1
These ripples on the surface of a coastal sand dune in New Jersey, USA, illustrate a common self-organization phenomenon: regularity of size and spacing of bedforms.

quences of bedforms of consistent shape and size are found in stream beds and on aeolian dunes. Polygonal patterns form in (especially) periglacial environments, and patterned ground is found in other environments as well. Remarkable visual and statistical regularity is found in the topological patterns of fluvial channel networks; and, indeed, in natural flow networks in general (Woldenberg 1968).

There is room for considerable doubt and debate as to the extent to which forms allow interpretation or prediction of processes or evolution. However, there is no doubt that form or geometry can be, and is, used in some situations to make inferences about the origins of the features. U-shaped valleys may form by other processes, but are generally diagnostic of alpine glacial erosion. Beach face slopes reflect wave energy dissipation regimes (Wright and Short 1984). Bending of tree trunks on alluvial floodplains records something of the history of flooding (Hupp 1988). The convexity or concavity of hillslope profiles reflects the relative importance of creep or wash transport (Carson and Kirkby 1972). The characteristic shapes of desert dunes reflect different combinations of wind regimes, sand supply, and vegetation cover (figure 3.1; McKee 1979).

Form–process relationships are not foolproof. In some cases, they may be misleading. For example, several different channel network development mechanisms, or assumptions about energy dissipation or network structure, can reproduce the characteristic geometry of dendritic drainage networks (Kirchner 1993). The success of any given concept or model in predicting form may, then, give false confidence about our ability to explain and interpret the resulting form. However, as a general principle, the presence of regular, repetitive, recurrent geometries in the landscape is evidence of some form of order in system structure, if not in evolution or process.

The soil-landscape paradigm

Much of my recent fieldwork has dealt with soil geomorphology in the south-eastern US coastal plain. That work has emphasized the complexities and spatial variability of the soil landscapes there, which are (as in many other soil landscapes) often quite dramatic. Yet, if you lead me to a given location within that region, I (or any other experienced field person) can readily predict, with very little information, the dominant soil which will occur there (see figure 3.2). True, the complexity and variability of the soil landscape may be such that the dominant soil type would only be found in half or less of the samples. However, the likelihood that the predicted soil will be found in more of the samples than any other soil type is high, and the probability that it will be somewhere at the site is even higher.

The fact is, despite the variability of the soil cover and the complexities of pedogenesis, there are systematic relationships be-

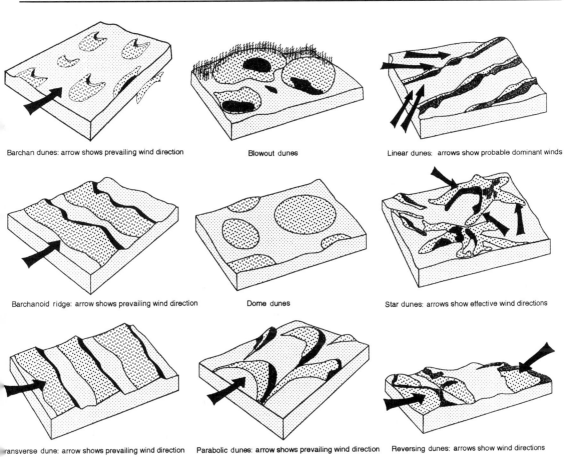

Barchan dunes: arrow shows prevailing wind direction   Blowout dunes   Linear dunes: arrows show probable dominant winds

Barchanoid ridge: arrow shows prevailing wind direction   Dome dunes   Star dunes: arrows show effective wind directions

Transverse dune: arrow shows prevailing wind direction   Parabolic dunes: arrow shows prevailing wind direction   Reversing dunes: arrows show wind directions

Figure 3.1
Form–process relationships. These common dune types (from McKee, 1979) reflect the interaction of wind climatology, sand supply, and vegetation. Arrows indicate the direction of the dominant, formative winds.

tween soils and landscape properties, such that soil types and soil properties can be predicted with considerable reliability based on observation of landscape properties such as topography, drainage, lithology, and vegetation. This "soil-landscape paradigm" is the operational underpinning of soil surveying and mapping, and the conceptual basis of much modern pedology (Hudson 1992; Gerrard 1993; Johnson and Hole 1994; McSweeney et al. 1994). A good soil map, and the fact that soils can even be mapped reasonably well without digging up the whole landscape, suggests considerable order in earth surface systems.

An orderly perspective

I have already argued, and will do so again, that the world is not a regular, ordered, sensible place whereby any observed disorder or

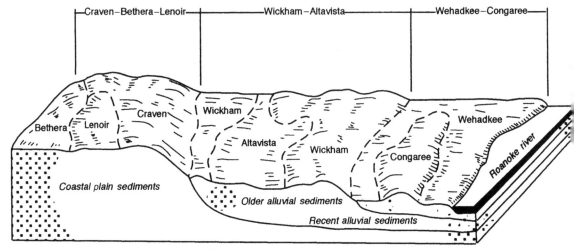

Figure 3.2
Soil–landscape relationships for a typical landscape unit in north-eastern North Carolina. The local spatial variability of soils is embedded within a broad-scale order typified by this pattern, where there are regular, systematic relationships between soil properties, topography, and parent materials (from "Soil Survey of Northhampton County, North Carolina," US Department of Agriculture, 1994).

complexity is merely awaiting our ability to perceive and unravel the underlying order. However, the world is also not a completely complex, complicated, random, chaotic place, where anything can happen any place, any time. Clearly, all combinations of all the variables are not possible. Beech trees will not grow in permafrost, barchans will not form in mud, and frost heave will not operate in the tropics. But, just as clearly, more than one outcome or one combination of variables is possible. Longleaf pine is likely to be growing in sand, but may grow in other soils, too; laterites are often associated with tropical weathering, but may form in extratropical climates; and stream aggradation could indicate a rise in base level, but could also signal an increased upstream sediment supply or tectonic adjustments.

When you start with the premise that anything could happen any time or any place, and then observe the far smaller number of things that actually do occur, it is clear that there is great order in nature in general, and in earth surface systems in particular. When you start with the premise that there is one possible outcome associated with a given history and/or set of environmental controls, and then observe the considerably greater number of outcomes that actually do occur, it is clear that nature in general and earth surface systems in particular are quite complex. The first part of this chapter, and most of human scientific endeavor, has focused on the order. We now turn our attention to disorder. Compared with the preceding section, it is going to be longer and more detailed, for two reasons. First, human comprehension and the scientific goal of explanation make it inevitable that examples of order in earth surface systems will be more intuitively obvious to us, and generalized examples suffice to make the point. Second, the traditional scientific focus on the orderly side of things makes a detailed review of order both unnecessary and impossibly daunting (at least for me!)

The Disorder of Things

The world is full of things we don't understand, haven't figured out yet, or that are just too imposing to get a handle on. That's not the kind of disorder I'll be discussing here. Rather, we will focus on complexity and variability in ESS that arises from the structure or functioning of the ESS themselves. This kind or disorder or complexity will be apparent whether we have figured it out or not. There are bound to be other ways to detect and assess this kind of disorder, but the map I have at hand is based on complex nonlinear systems.

Complex nonlinear dynamics in nature

Not all earth surface systems are nonlinear. But many are. Not all nonlinear systems exhibit instability, chaos, and self-organization. But some do. Not all nonlinear systems which can exhibit deterministic complexity do so at all times or under all circumstances. Having said all that, let it be known that the material that follows is *not* an effort to convince anyone that the whole world is nonlinear, unstable, chaotic, and self-organizing. I have focused on systems and situations where deterministic complexity has been observed to make the case that such behavior is quite common, and something we have to deal with in attempting to understand ESS.

Early work in NDS theory, and much current work, is focused on mathematical models. This body of work continues to be quite useful and influential, and is necessary to inform field and laboratory studies of complex nonlinear ESS. But, ultimately, we must confront the issue of the extent to which deterministic complexity is a property of models and equations (a mathematical artifact) as opposed to something observed in the real world. Further, field scientists may be largely unimpressed by conclusions drawn from equation systems or simulation models, and even the most devoted modeler eventually aspires to test the model against nature.

Therefore, with only a bit of further ado, there follows a whirlwind tour of a body of work showing evidence of deterministic complexity in the real world. The latter is here interpreted broadly to include field studies, laboratory studies, and analyses of real-world data sets, as opposed to research based strictly on models or abstractions. Results of modeling studies are included only when they are either derived from and used to explain real-world data or successfully tested against real-world evidence. I urge you to consult the original sources to indulge your curiosities and skepticisms!

Chaos

Several studies have explicitly addressed deterministic chaos in time series or spatial data from hydrologic records, tree rings, and topographic images. Jayawardena and Lai (1994) used a time-delay embedding approach for reconstructing phase-space diagrams for time series of daily rainfall and streamflow data for Hong Kong. Results suggest that these hydrologic time series are characterized by a chaotic attractor. Similarly, analysis of 1847–1992 volume records of the Great Salt Lake, Utah suggests the presence of a low-dimensional strange attractor (Sangoyomi et al. 1996). Red Spruce decline in the north-eastern US is believed to be due to one or more of several factors: climate forcings, acid deposition, and stand maturity. Ring widths show complex time series with regional order, but with much highly irregular, unexplained variability. Van Deusen (1990) found that the statistical models which best fit the ring width data can exhibit deterministic chaos. With regard to spatial data, Rubin (1992) developed a method for distinguishing chaos, noise, and periodicity in spatial patterns and applied it to images of aeolian sand ripples. The ripples exhibit downcurrent coupling, sensitivity to initial conditions, self-organization, and decaying predictability with forecasting distance. Nonperiodic, deterministic nonlinear interactions between ripples, rather than quasiperiodicity or randomness, appears to be responsible for the observed patterns (Rubin 1992).

A number of other studies have identified chaotic behavior in models of earth surface systems, and have verified the model predictions with field evidence. Models describing pedogenesis of Ultisols in eastern North Carolina are unstable and potentially chaotic, and predict that soils on older surfaces should be more diverse, even where other soil-forming factors are negligibly variable. Phillips (1993a) confirmed the prediction by comparing the taxonomic diversity of soils at two adjacent sites with the same parent material, climate, biota, and topography, but differing in age. Phillips (1992c) also found that the interactions between soil moisture, infiltration, and runoff are unstable and potentially chaotic for infiltration-excess overland flow (and stable for saturation-excess overland flow). He used data from runoff plot experiments where infiltration-excess flow was generated to demonstrate the sensitivity to initial conditions predicted by the model (figure 3.3). The same general approach has been applied to at-a-station hydraulic geometry in stream channels. The Darcy–Weisbach equation, reformulated as a nonlinear dynamical system showing velocity, width–depth, slope, and roughness as functions of each other, is unstable and potentially chaotic. This is manifested as multiple modes of adjustment, whereby the hydraulic variables may change in

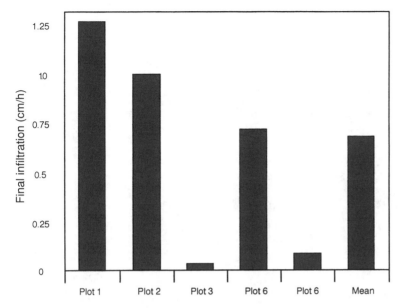

Figure 3.3
*Chaos and sensitivity to
initial conditions in
infiltration and runoff.
Final infiltration capacities
in the plot experiments of
Ewing and Mitchell (1986)
show considerable
variation, despite the
uniformity of the plots and
of simulated rainfall
applications. This result
was interpreted by Phillips
(1992c) as chaotic
sensitivity to minor,
unmeasured variations in
plot characteristics or
simulated rainfall rates.*

opposite-from-expected directions in response to changes in discharge. Data from surface runoff experiments and stream channel
hydraulic geometry exhibit this behavior (Phillips 1990d, 1992c).

Evidence of chaos has also been found in hydrothermal eruptions. A state space reconstruction of Old Faithful geyser (Wyo-
ming) eruptions by Nicholls et al. (1994) shows a chaotic pattern of
intervals between eruptions, and inability to predict more than one
event into the future. Conversely, predictions one event into the
future are highly reliable, indicating a deterministically rather than
a stochastically complex system. The behavior is consistent with the
nonlinear equations governing hydrothermal geysers.

Finally, the generally accepted qualitative understanding of
relationships between sea-level rise, hydroperiods, organic and
inorganic sedimentation, vegetation, and erosion in salt marshes
indicates a complex system which can be shown to be unstable and
chaotic. This model, adapted by Phillips (1992e) from Reed (1990),
predicts increasingly complex marsh shoreline configurations and
spatial mosaics of marsh types and open water as coastal submergence proceeds. Results for Delaware Bay, New Jersey, based on
historical aerial photographs, show an increasingly complex shoreline (as indicated by the fractal dimension) as erosion and drowning
proceed (plate 3.2; Phillips 1992e, 1989g).

Instability

The terms stability and instability are used in a variety of ways in
the literature. There are numerous studies indicating unstable be-

Plate 3.2
Aerial photograph of a portion of the Delaware Bay, New Jersey, USA, coastline, showing estuarine salt marshes. A nonlinear dynamical systems model of marsh response to sea-level rise indicates unstable, chaotic behavior, and predicts an increase in marsh shoreline irregularity (as measured by the fractal dimension) over time. The latter prediction has been confirmed by historical aerial photographs.

havior in ESS which might (or might not) be interpreted as dynamical instability in the sense that the latter term is used in this book. But even when we restrict ourselves to explicit consideration of dynamical systems instability and sensitivity to small perturbations, we are left with plenty of real-world evidence. Take river morphology. Coastal rivers of New South Wales, Australia show evidence of fluctuation between two or more morphological states (based on the balance between channel and floodplain erosion and deposition), rather than equilibrium behavior (Nanson and Erskine 1988), and are sensitive to small perturbations to the erosion/deposition ratio. Rhoads and Welford (1991) have explained the initiation of river meanders on the basis of bar-bend theory, involving growth and migration of infinitesimal flow perturbations (figure 3.4). Carson and LaPointe's (1983) analysis of 15 rivers demonstrates the inherent asymmetry of meander planforms. They attribute it, based on field data, to the persistence and growth of helical flow perturbations and downchannel persistence of cross-sectional velocity distributions well past the bend that forms them. At a longer timescale, Knox (1985, 1993) used dimensions

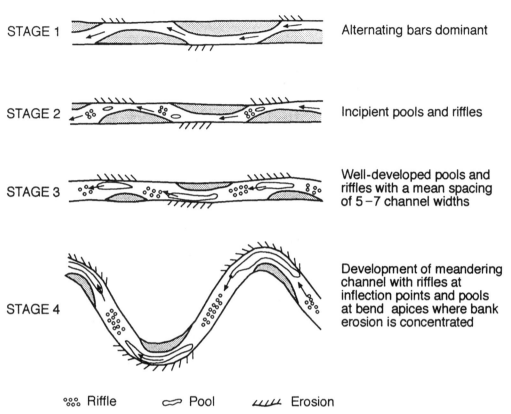

STAGE 1 — Alternating bars dominant

STAGE 2 — Incipient pools and riffles

STAGE 3 — Well-developed pools and riffles with a mean spacing of 5–7 channel widths

STAGE 4 — Development of meandering channel with riffles at inflection points and pools at bend apices where bank erosion is concentrated

°°° Riffle   ⊂⊃ Pool   ⌐⌐⌐⌐ Erosion

Figure 3.4
The classic model showing the transformation of a straight into a meandering channel (from Knighton 1984). Evidence suggests that the meandering process is initiated by the unstable growth of small flow perturbations.

of Holocene relict channels and sedimentology of point bars to reconstruct the chronology of floods in the Upper Mississippi Valley, Wisconsin. Fluvial responses are associated with long-term episodic mobility and storage of sediment. They are disproportionately large compared to climate changes, indicating an unstable response.

Flow phenomena also exhibit instability, even at levels of aggregation far above the well-known complexities of turbulence. Field experiments in Kenya confirmed a considerable body of previous theoretical assertions that sheetflow is unstable, leading to separation into rills (Dunne and Aubry 1986). Parsons et al. (1996) examined changes in inter-rill runoff and erosion associated with shrub invasion of grasslands in southern Arizona. They found that a minor environmental change in vegetation cover led to disproportionately large geomorphic changes in terms of runoff rates and velocities, and soil erosion. In channels, at a given cross-section velocity, width–depth, slope, and roughness often exhibit multiple modes of adjustment, whereby the hydraulic variables may change in opposite-from-expected directions in response to changes in discharge. Data from the Bogue Phalia River, Mississippi exhibits this behavior, which is explained by the unstable relationship among the hydraulic variables in standard hydraulic equations (Phillips 1990d, 1991d).

Instability has also been observed in the evolution of soils and landforms. Pedogenesis on podzolized soils in Australia after sand mining shows that rapid leaching leads to thicker E-horizons than existed pre-mining, due to destruction of indurated B-horizons. This supports the notion of B-horizon formation as a negative feedback on depth-to-B, and indicates instability of the soil profile to disturbance (Prosser 'and Roseby 1995). In some regions of the Southern Alps, New Zealand, soil patterns and stratigraphy indicate continuous instability and a complex spatial pattern of soil development and bedrock outcrops. Initial erosion of the soil mantle is from bedrock hollows, and the further instability propagates from there. During this phase a complex array of soil profile forms results (Tonkin and Basher 1990).

Other ESS phenomena have also been shown to be unstable. Brinkmann (1989) found that hydrologic and biogeochemical cycles of a *terre firme* lowland forest in Brazil were in a steady-state equilibrium, but one highly sensitive to small changes and therefore unstable. In Mali, vegetation removal and trampling initiates water and wind erosion in fossil dunes. Once initiated, the process is characterized by unstable self-acceleration (Barth 1982).

As with chaotic behavior, in some cases modeling studies have found dynamical instabilities in ESS models, and verified model predictions with field observations or data. This is the case for Liu's (1995) work on Milankovitch glacial cycles. Dynamical instabilities can explain glacial cycles, due to the extreme sensitivity of Milankovitch cycles to changes in obliquity frequency. Theoretical predictions and model results show good agreement with the late Pleistocene ^{18}O record (Liu 1995). A nonlinear model with unstable relationships suggests that multiple equilibria associated with nonlinear feedbacks between climate, sea level, subsidence, and sedimentation could lead to large shifts in coastal onlaps from small perturbations. The model explains anomalous cyclicity frequently observed in onlap stratigraphy (Gaffin and Maasch 1991).

A model accounting for feedback between precipitation and development of a root zone was used to examine the response of local hydrologic cycles to global temperature change. Instabilities lead to differential response and divergence into tropical forest, savannah, semi-arid, and desert environments from similar initial conditions. Results are supported by paleoclimatic data for northern Africa (Lapenis and Shabalova 1994). Rhoads (1988) developed a statistical causal model from field data. The model shows that mutual adjustments between form and process in a desert mountain fluvial system in Arizona are meta-stable (plate 3.3). When displaced by

Plate 3.3
An ephemeral desert mountain stream in Maricopa County, Arizona, USA. Rhoads (1988)
showed that these fluvial systems, when disturbed, do not return to their pre-disturbance state.

exogenous disturbances, the system does not return to or fluctuate around initial values, but moves to a new state.

Scheidegger (1983) proposed a general instability principle for geomorphology based on a combination of fundamental concepts of both systems theory and geomorphology. According to the instability principle, topographic variations persist or grow over some finite time, increasing the topographic complexity of the landscape. Scheidegger supports his proposal with several field examples from hillslopes, valley cross-sections, and rivers in Europe.

Increasing variability over time

If an ESS is unstable, chaotic, or self-organizing, it should exhibit increasing spatial complexity over time, in the absence of (or independently of) any changes in external controls or influences. While changes in spatial complexity are relatively easy to ascertain, establishing that changes in external controls have been negligible, or that the observed changes are independent of them, is quite difficult. Despite this challenge, several studies have demonstrated patterns of spatial variability which become increasingly complex or less uniform over time, due entirely to the dynamics of the system itself rather than to external forcings.

Dryland degradation (desertification) apparently involves increasing variability over time in the form of a progressively more complex pattern of vegetation and soil nutrient resources. Feedbacks linking climate change, human interactions, precipitation, evapotranspiration, soil temperature and moisture, soil organic nitrogen, erosion, and runoff are important in desertification processes. Divergence in the form of concentrations of resources (soil nutrients and moisture) and resource losses is illustrated via field data from New Mexico (Schlesinger et al. 1990). Additional field data from the south-western US show that the spatial patterns of soil nutrients exhibit increasing local variability as shrubs invade semi-arid grasslands, which typically accompanies overgrazing (Schlesinger et al. 1996). Working in the grasslands/shrublands of Walnut Gulch, Arizona, Abrahams and others (1995) found that grass-to-shrub vegetation change increases inter-rill erosion and runoff, and further increases the spatial heterogeneity of the plant canopy. This leads to additional variability in desert pavement formation, soil fertility, erosion, and hydrologic response. Puigdefábregas and Sánchez (1996) also found that differences between bare ground and vegetated patches in Spain increase over time due to self-reinforcing mechanisms. Their field studies demonstrate that the primary mechanisms for increasing variability are associated with water and sediment circulation.

In the Chaparral of southern California, microscale soil heterogeneity appears to be a result rather than a cause of vegetation heterogeneity. This indicates that plants progressively modify soil, increasing soil variability over time (Beatty 1987). Soil variability can also increase over longer time scales. Despite uniform parent material, spatial variability in the degree of soil development increased over time in a chronosequence of 24 pedons on sandy lake terraces in Michigan, with ages of 3,000 to 11,000 years BP (Barrett and Schaetzl 1993). In the parabolic dune systems of eastern Australia, as the age of dunes and associated soils increases, not only the mean thickness, but the variability of the dimensions of E- and B-horizons increases (figure 3.5; Thompson 1983; Thompson and Bowman 1984).

Divergence over time – spatial differentiation of the landscape – also occurs in salt marshes, particularly those stressed by coastal submergence. In eroding and drowning marshes in Lousiana, "broken" marsh expands at the expense of unbroken marsh due to erosion processes, increasing the spatial variability of the marsh/open water mosaic (Nyman et al. 1994). Depressions associated with localized vegetation suppression grow unstably in marshes of south-eastern Australia, and there is an increasing density of salt pans (and thus increasing irregularity of the marsh surface) with marsh age (Boston 1983).

Dune systems

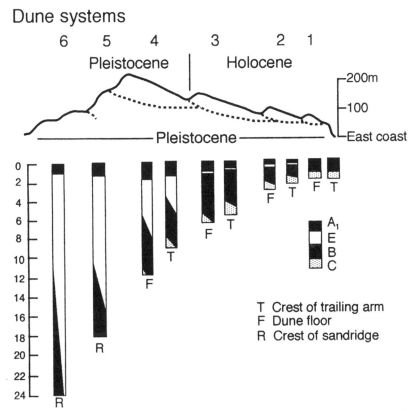

Figure 3.5
A chronosequence of dune soils from Australia, showing not only the development of podzolized soils, but the increasing variability of E- and B-horizon thickness over time (from Thompson 1983). Dune systems become progressively older from 1 to 6. The lengths of the diagonal lines indicate the variability in horizon depths and thicknesses.

Spatial variability in the form of topographic roughness may sometimes increase over time even when external controls are constant. Roughness at 15 sites on deposits of known age in north Norway was measured by Elliott (1989) via fractal dimensions. Despite apparent constancy in other factors, over 40 years the surfaces become rougher and more variable.

At the even shorter time scales relevant to hydrologic stystems, the area under a single large conifer in Vermont was found by Hardy and Albert (1995) to accumulate 34 percent less snow during the peak snow accumulation period than surrounding unaffected areas. Following the peak ablation season, snow cover under the tree was 18 percent of the unaffected zone. The enhanced difference is attributed to lower temperatures at the snow–ground interface associated with the reduced snow cover under the tree, which led to stronger thermal gradients and increased rates of heat flux compared with the unaffected treeless area. Anderson and Cassell's (1986) measurements of variablity of soil physical and hydraulic properties in eastern North Carolina show that variability of hydraulic conductivity increases to a maximum at saturation, implying increasingly complex flow as the soil wets during a single event.

Self-organization

The several faces of self-organization, and their fundamental similarities, have already been discussed, along with the links to instability and deterministic chaos. Many examples of self-organization, such as a sequence of ripples on a lake bottom, may appear to be quite the opposite of disorder, given their geometric regularities. However, in some cases self-organized patterns have been shown to arise from complex nonlinear dynamics. Self-organization is also explicitly linked with chaos and instability in a formal sense, as shown in chapter 2.

Much work has focused on models, but some models have been linked to or tested with field evidence. Ahnert's (1994) simulation model shows that non-periglacial sorted nets can develop spontaneously on any piece of unobstructed ground with little or no slope if it bears a loose, discontinuous cover of pebbles, and if these can move in small steps with equal probability in all directions. Physical causes of movement are irrelevant. Field examples of non-periglacial sorted nets are provided from Germany, and from the results of lab experiments. Dunne et al. (1991), using a physical model verified in the field, show self-organization in the form of a systematic hydrologic response at the hillslope scale arising from the mechanics of infiltration and runoff, which are highly sensitive to local effects of vegetation and microtopography.

Spontaneous pattern formation and autogenic processes have been observed in a number of ESS, notably with respect to interactions of water flows and their bedforms. Nelson's (1990) flume studies showed that topographically induced steering in straight channels creates instabilities in formation of alternate bars, with these initial instabilities leading to eventual broader-scale stability. In streams of south-west England, a model of riffle-pool sequence formation as an autogenic process was developed to explain field observations (Clifford 1993). Local scour creates deposition downstream, which then generates the next downstream flow irregularity. The form–process feedbacks create a system with both deterministic and (apparently) stochastic elements. Chin (1989) found self-organization in the regularity and rhythmicity of step-pool sequences in the Santa Monica mountains, California, and other mountain streams (plate 3.4). In coastal and fluvial settings, sand bars formed by partially standing waves often have regular spacings equal to half the surface wave length. Rey et al. (1995) reproduced these features in a wave tank, showing the self-organized development of regular wave features from an initially flattened surface, and the initial growth of instabilities (small ripples) followed by finite-amplitude stability.

Landscape-scale self-organization has been observed with respect to desert regolith development, drainage basin sediment systems, and erosion–deposition patterns. The mosaic of bedrock outcrops and soil cover in the Negev desert leads to an increasing divergence of regolith development. Runoff from rock areas concentrates

Plate 3.4
Step-pool sequence from the Santa Monica Mountains, California, USA. The energetics of step-pool sequences indicate a self-organizing process (Chin, 1989; photograph courtesy of Anne Chin).

water, infiltration, and leaching in soil-covered areas, and leads to salinity concentrations, resulting in an increasingly organized pattern of bedrock outcrops and regolith covers (Yair 1990). Engelen and Venneker (1988) argue, partly on the basis of field surveys of the Upper Boite river, Italy, that self-organization is indicated in the organization of drainage basin sediment systems into erosion–transport–accumulation chains. Similarly, the simultaneous presence of order and randomness is indicated by alluvial landscapes in arid central Australia. The landscape is self-organized into scour–transport–fill sequences, but these are complex enough to be modeled as a random field (Pickup and Chewings 1986).

The mutual influences of vegetation and substrate characteristics can also result in self-organization. In salt marshes of south-eastern North Carolina, new substrates exhibit gradients of elevation, hydroperiod, salinity, and redox potential. This creates a continuum of overlapping vegetation habitats. Once vegetation establishes itself, the marsh vegetation alters the chemical and physical nature of the substrate, creating conditions favorable for itself and sometimes unfavorable for competing species. The orginal overlapping continuum of habitats is transformed into a well-organized spatial pattern of vegetation and associated hydrogeomorphic characteristics with well-defined boundaries (Hackney et al. 1996).

Most of the studies cited above describe clear cases of spontaneous, autogenic self-organization, but in the whole do not make explicit reference to nonlinear dynamical systems and related conceptual frameworks. There are other studies which do draw such explicit links. Hallet (1990) demonstrates self-organization, and discusses the underlying processes, in pattern formation in freezing

Plate 3.5
A ferricrete pseudomorph from Craven County, North Carolina, USA. Nahon (1991) explained the formation of ferruginous nodules and iron crusts in terms of geochemical self-organization.

soils in Alaska and Canada. Orford et al. (1991) interpreted his field evidence from Ireland and Canada to show that gravel barriers can be described as a self-organizing system with two stable attractors, corresponding to swash- and drift-alignments. At a more detailed scale, drift-aligned gravel barriers in Ireland and Canada self-organize at two levels: clast selection and sorting, and planform. Evolution tends toward maximum-entropy low-organization or minimum-entropy high-organization arrangements (Carter and Orford 1991).

Haigh (1989) presented evidence that the evolution of an anthropogenic gully system in Arizona can be successfully described by a catastrophe theory model based on instability with respect to transitions between two attractors. He also demonstrated that the systematic rank-size distribution of Himalayan landslides is predicted by principles of self-organization in dissipative systems (Haigh 1988). Lateritic weathering may also be self-organizing. Formation of ferruginous nodular horizons, and of conglomeratic iron crusts, can be explained in terms of autonomous changes in soil geochemical processes without the influence of external factors such as climate (plate 3.5; Nahon 1991).

Divergence from similar initial conditions

In the geosciences initial conditions are often unknown, so that often the emphasis is on increasing spatial variability over time, or on sensitivity to small perturbations, rather than on sensitivity to initial conditions. There is substantial real-world evidence, however, that shows earth surface systems where there is clear divergence from the same, or very similar, initial conditions.

Much of the evidence is from studies of the interactions between vegetation and environmental factors such as microclimate, hydrology, and soils. Many examples are given in Wilson and Agnew's (1992) extensive review. For example, fog precipitation occurs when taller vegetation traps droplets from low clouds or fog. This enhances water input to tall vegetation, compared to adjacent low plant cover, and favors growth of taller vegetation. Sharply bounded communities of tall/short (for example, trees versus grass) vegetation develop from an initial condition of a mixed community. Field evidence is abundant in the form of sharp or sharpened boundaries in forest patches amid grassland in Kenya, the Carribean coast of South America, the Ecuadorean Andes, and California; and in enhancement of water input and accompanying plant response in taller vegetation in Vermont mountains, Chilean forests, tussock grasslands in New Zealand, and Monterey dunes, California. Numerous examples are cited by Wilson and Agnew (1992). Tall vegetation also sometimes reduces fluctuations in temperature near the soil surface. This can favor growth or re-establishment of the same species if they are less cold tolerant. Divergence occurs as sharpened boundaries and as persistence of disturbance-induced changes in treeline in New Zealand and Alaskan tundra (Wilson and Agnew 1992).

Vegetation locally enhances runoff infiltration. This increases water availability to plants, which "trap" more runoff and further concentrate moisture and vegetation. This occurs mainly in drylands. Divergence in the form of the development of patchy vegetation bare ground mosaics from orginally more uniform patterns has been observed in Serengeti grasslands, East Africa, semi-arid eastern Australia, and the Chihuahua desert, Mexico (Wilson and Agnew 1992).

In the Norfolk marshes, England, vegetated marsh areas trap more sediment than non-vegetated ones, which further reduces indundation and lowers (or prevents the rise of) salinity. Depressions trap more water, which increases salinity and inhibits vegetation. Destruction of vegetation eliminates trapping functions and may initiate panne formation (Yapp and Johns 1917, cited in Wilson and Agnew 1992; Pethick 1974). Similar processes occur in south-eastern Australian marshes (Boston 1983). In some marshes in Lousiana positive feedback reinforces marsh loss. Flooding stress on plants limits vertical accretion, which further increases flooding and decreases plant production (Nyman et al. 1993). In other, nearby marshes physical erosion is the major marsh loss mechanism, but similar positive reinforcement is evident (Nyman et al. 1994). In all cases an originally more-or-less uniform marsh surface is transformed into an increasingly variable mosaic of vegetation patches and geomorphic micro-environments.

Acid-tolerant vegetation produces acidic litter or roots, which in turn further depress soil pH, favoring low base status vegetation and inhibiting other types, sometimes mediated through the effects of pH on earthworms or other soil fauna. Examples of coevolutionary divergence in soil and vegetation in this manner in Derbyshire, England, are cited by Wilson and Agnew (1992). Plants may also increase or decrease soil nutrients, which favors those plants which either respond better to increases or better tolerate the decreases. Field evidence is in the form of development and persistence of vegetation/soil nutrient patches from previously more homo-geneous landscapes. Wilson and Agnew (1992) give examples from arid Australia, mid-western US, Alaskan forests, and bogs in Minnesota and northern England. In other instances, plants salinify the soil through their litter. These species are tolerant of the higher salinity, and therefore develop patches. Evidence comes from Cohune Palm stands in Belize, and arid shrublands in the Karakum desert and in the western US (Wilson and Agnew 1992).

Grazing disturbance can foster divergence by leading to alternative multiple stable states of vegetation communities and environmental factors in rangelands, as shown by work in the Middle East, the western US and Australia (Westoby et al. 1989; Freidel 1991; Laycock 1991; Tausch et al. 1993; Walker 1993; Hobbs 1994). Balling's (1988, 1989) studies along the Mexico–US border show how vegetation discontinuities due to overgrazing create climatic variations, which reinforce the vegetation patterns via microclimate and soil hydrology. In areas where aeolian erosion is prevalent, vegetation can trap blowing sediment. The deeper soil encourages plant growth, further intensifying the effect. This phenomenon is evident in the development of shrub patches in southern Australian rangeland, Iraqi deserts, New Zealand mountains, and western US rangeland (Wilson and Agnew 1992). Frequent fire favors the establishment of fire-tolerant species which produce flammable litter. Closed forest suppresses flammable-litter vegetation, which decreases fire frequency. Thus a uniform mosaic of forest and savannah shows local divergence into sharp savannah/closed forest boundaries in Australia and east Africa (Wilson and Agnew 1992).

While divergence from similar or uniform initial conditions is most readily observed in relatively short-lived or fast-changing phenomena, such as vegetation and soil chemistry, divergence has also been noted over geological time scales. For example, Twidale (1991) describes a conceptual model whereby differences in weathering and erosion susceptibility are reinforced by positive feedbacks, leading to increasing and diverging relief. This model explains the persistence of relief increases in a variety of tectonic settings over 60–100 million years. Seven Australian landscapes where field evidence is consistent with the model are described (Twidale 1991). McDonald and Busacca (1990) found that variable rates of loess deposition contributed to dramatic spatial divergence from identical starting points in a soil stratigraphic unit in the channeled scablands of eastern Washington. Immediately downwind of a major loess source, paleosol development bifurcates into two distinct soils, apparently due to interactions between loess deposition and formation of superimposed calcic soils.

Sensitivity to initial conditions

Chaos and instability imply sensitivity to initial conditions. Several studies, while not explicitly addressing chaos, stability, or nonlinear

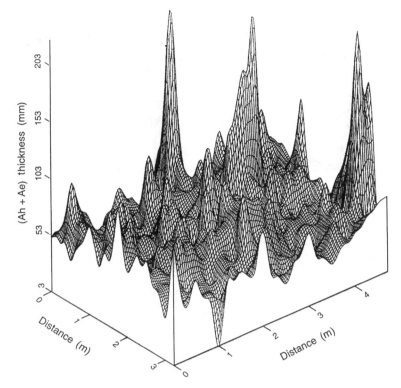

Figure 3.6
Spatial variability in A-horizon thickness. Price (1994) found that microtopographic variations led to preferential nodes of infiltration and to disproportionately large variations in horizon thicknesses in Canadian podzols.

dynamics, have none the less effectively demonstrated ESS where slight variations in initial conditions result in much larger variations as the system evolves. In a podzolized forest soil in Canada, Price (1994) showed how microtopographic variations led to preferential nodes of infiltration and "funnel" effects, which subsequently created disproportionately large variations in the thickness of A and B soil horizons (figure 3.6). Spatial variability of soil properties in semi-arid Botswana is likewise related to microtopography and is also disproportionately high relative to microtopographic variation, suggesting sensitivity of soil properties to minor initial variations (Miller et al. 1994). Microtopography and tree islands in the upper timberline zone of Colorado influence near-surface winds, which in turn affect the distribution of snow cover. This affects soil processes and properties, and leads to a patchy pattern of vegetation, soil, and microclimate attributable to the sensitivity to minor initial variations in topography and vegetation cover (Holtmeier and Broll 1992).

Sensitivity to initial conditions is also evident in "fingered flow" phenomena in homogeneous sandy soils (figure 3.7). Flow fingers in homogeneous coarse-grained soils form from instabilities in the wetting front. The initial water content in and around the fingers, which tend to persist for long periods, is crucial (Hill and Parlange 1972; Hillel and Baker 1988; Ritsema et al. 1993; Liu et al. 1994).

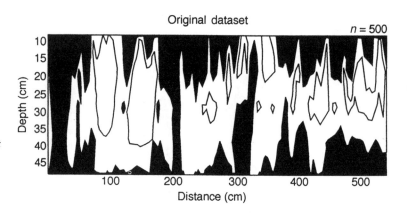

Figure 3.7
*"Fingered flow" is
illustrated by this contour
plot showing water content
of a uniform sand
subjected to uniform water
application rates (Ritsema
and Dekker, 1995).*

Bare, homogeneous dune sands in Mexico and New Mexico
show highly irregular soil moisture patterns. Small differences in
water content lead to large differences in hydraulic conductivity,
and the anisotropy in water content and conductivity tends to be
magnified over time. Preferential flow also occurs (Ritsema and
Dekker 1994). Ritsema and Dekker (1995) found that incoming
water in the upper (~1 cm) layer in water-repellent soils is deflected
from vertical paths toward places where vertical infiltration
dominates. Microrelief is critical. Spatially concentrated water/
solutes create unstable wetting fronts. Experiments on level soil,
spodosol and sandy entisols show differences of 100–400 percent in
local water input even where the rate of surface water application
is uniform. Small variations in soil moisture at the onset of
a precipitation event can foster large variations in hydrologic
response at broader scales, too. Troch et al. (1994) found that the
flood frequency distribution of a small basin in the Pennsylvania
Appalachians is insensitive to maximum rainfall intensity, but
strongly affected by initial conditions of soil moisture storage
capacity.

The relationships between landform morphology and processes
can also exhibit sensitivity to initial conditions. Allen (1974) shows
how the singularity of beach environments results in unique
longshore current developments at Sandy Hook, New Jersey.
Vertical accretion on tidal flats, and subsequent salt marsh development, is sensitive to the distributions and grazing predation of
benthic microalgae in the initial stages of mudflat development
(Coles 1979). Salt marsh development in Massachusetts and freshwater tidal marsh development in New Jersey shows the influence
of individual perturbation events and historical persistence. The
structure of plant communities at any point in time depends on,
among other things, historical geomorphic development trends.
Similar implications were found in a paleoecological study of
Holocene tidal freshwater marsh in the upper Delaware river
estuary (Orson and Howes 1992; Orson et al. 1992).

Increasing K-entropy

Increases in Kolmogorov (K-) entropy are linked to unstable, chaotic, self-organizing evolution, as described in chapter 2. The broad spatial and temporal scales of landscape evolution are somewhat more amenable to this approach than to other means of assessing the possibility of deterministic complexity. Topographic maps of Europe show higher entropy with greater relief, and lower entropy with smoother relief (these trends refer to K-entropy: opposite of the entropy calculated by authors Zdenkovic and Scheidegger 1989). Thus increases or decreases in relief seem to correspond with changes in K-entropy. Research on the Iberian peninsula suggests that fluvial landscape dissection expands pedo- and ecospheres, allowing for self-organization. Topographic maps of six surfaces ranging in age from 2.5 million years to modern, representing incision into an older surface, were examined for measures of complexity, including K-entropy. As incision proceeds, complexity/entropy increases (Ibanez 1994). Earlier work on the Iberian peninsula, using basin order as a surrogate for stages of dissection, investigated the relationship between fluvial dissection and the diversity of soil cover (Ibanez et al. 1990). As erosion advanced in the Quaternary, soil landscapes became enriched and diversified (entropy increase), based on mapping of drainage networks and incision from air photos, field study and mapping of soils, and calculation of an entropy-based diversity index. As basin order increases, so does soil entropy. Fiorentino et al. (1992) also found entropy in network spatial structure to increase with basin order and magnitude in a theoretical model confirmed by data from three basins in southern Italy. Relief-reducing degradation in soil-covered landscapes of southern Britain reduces fractal dimensions and K-entropy (Culling 1987; Culling and Datko 1987). Presumably, relief-increasing evolution should have the opposite effect, and K-entropy is an appropriate measure of the "chaoticity" of the process (Culling 1988b).

Complex behavior explained by nonlinear dynamics

Finally, there are a number of ESS where deterministic complexity is not necessarily demonstrated by data or field observations. Rather, the observed real-world behavior of the ESS is not readily explained *except* by NDS-based explanations. Pleistocene climate variability and ice sheet–bedrock feedbacks are a good example. The ^{18}O record at 100,000 years contains unexplained variability. Nonlinear coupling and internal variability arising from feedbacks between ice sheets and bedrock may account for the variability. Model results are consistent with three ocean ^{18}O records, and explain the trends better than alternative explanations (Birchfield and Ghil 1993). Similarly, land surface–atmosphere interactions and nonlinear feedbacks explain, at least in part, the per-

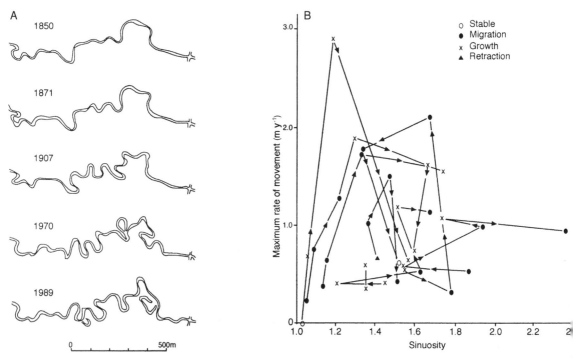

Figure 3.8
River planform change in England. (A), The sequences of change on the River Bollin near Wilmslow in north-west England; (B), rates of movement plotted against sinuousity for individual bends (from Hooke and Redmond 1992).

sistence of wet or dry conditions in the Sahel (Lare and Nicholson 1994).

Phase space plots of planform change in British rivers shows evidence of complex, nonlinear behavior. Nonlinear dynamics and positive feedbacks are plausible explanations of aspects of the observed behavior which cannot readily be linked to process–response mechanisms (figure 3.8; Hooke and Redmond 1992). Likewise, complex stratigraphy in Australian floodplains can be explained as the chaotic outcome of a few simple deterministic mechanisms, involving micro-, meso-, and macro-scale flood impacts (Bourke 1994). Fractal dimensions of real channel networks and fluvially eroded terrain are often scale-dependent, and have been the topic of much research and speculation. The scale transition is associated with drainage density, which can be interpreted as the transition between scales where unstable channel-forming flow processes yield to stable diffusive processes (Tarboton et al. 1992).

In the surf zone at Duck, North Carolina, nonlinear models gave better forecasts than linear ones of sediment suspension as a function of velocity. Including an "optimum" velocity history to incorporate persistence effects produced better results with both linear and nonlinear models. The complex, still not fully explained variation in suspended surf zone sediment indicates a deterministic nonlinear structure and a stochastic component (Jaffe and Rubin 1996).

In debris flows, Straub (1996) shows that the linear relationship between drop height and travel distance is explained by a dynamic model of sheared granular flows characterized by interparticle collisions. The model shows that at the microstructural level, granular flows are governed by deterministic chaos. At the bulk level, self-organization associated with the attractor controls the energy dissipation. The work also produced an empirically testable hypothesis: that rapid granular flow is the major flow regime in debris flows.

Unexpectedly weak and cyclical relationships between moisture surplus and runoff (as indicated by low correlations) in an eastern North Carolina basin with an extensive artificial channel system can be explained by the unstable relationships between moisture surplus, drainage density, channel efficiency, and runoff (Phillips and Steila 1984). Temporal trends of stream solute transport in Germany are complex and irregular, and do not correspond well with hydrologic or geochemical controls. Phase space portraits of calcium concentrations in successive events show no repeated points, and suggest a strange attractor which plausibly explains calcium runoff patterns (Kempel-Eggenberger 1993).

Deterministic uncertainty is a possible explanation of observed irregularities in relationships between surface horizon and parent material textures in soils. Data from eastern North Carolina show that deterministic uncertainty associated with depositional environments of parent material accounts for about 46 percent of the observed entropy in surface textures (Phillips 1994). Dekker and Ritsema (1994) were able to explain sand columns and other features in German and Dutch coastal sands as a result of unstable fingered flow, which occurs even in homogeneous dune sands.

Major forcing factors for Holocene forests of southern Ontario are climate, soil development, and internal forest dynamics. Interactions between individuals, populations, and the physical environment ensure that deterministic prediction of change will always be difficult. The complex behavior can be interpreted by viewing forests as nonlinear dynamical systems, and in southern Ontario they are shown to be heavily influenced by historical events (Bennett 1993). Similarly, instability to perturbations in wind-velocity fields can explain the formation and maintenance of wind-shaped wave forests in Canada (Robertson 1994).

Where to Now?

Anything is not possible, at least not in all places and at all times. Therefore, there is considerable order in earth surface systems. However, many things are possible and there is considerable disorder in ESS, much of it inherent in the structure and dynamics of the systems themselves. If I have done my job reasonably well, it will be clear by now that both order and complexity are common in the environment. They are often found at the same time, in the same place, in a system. Note that some of the concrete examples, such as landscape geometry and runoff response, appear in both the order and complexity sections of this chapter.

There are at least four different ways to interpret the simultaneous, intertwining presence of order and complexity in ESS. First, perhaps complexity is an artifact of spatial or temporal resolution. Once the "correct" resolution is found for studying a problem or analyzing a data set, the order becomes apparent and the complexity is merely a byproduct of inefficient or inappropriate lumping or splitting. A second, related view is that regularity and order dominate in ESS; complexities are superimposed on top of the orderly relationships and may obscure them. These first two views encompass the traditional view of environmental heterogeneity and spatial complexity. A third, diametrically opposed view is that complexity and irregularity are the norm in the structure and function of ESS. Regularity and order are mostly apparent, and are an artifact of lumping and aggregation which obscures the complex details. Finally, the view explored – and maybe, ultimately, espoused – here is that many ESS are inherently unstable systems which by their nature have both complex disorder and emerging, stable order at different scales.

None of the four viewpoints above is necessarily right or wrong; all have been operationally successful in helping scientists understand earth surface processes and features. However, I will argue that the fourth perspective – that ESS are frequently unstable – has the potential to unify the several threads of thought.

In the next chapter we will examine highly generalized, prototype ESS with the goal of demonstrating that:

- A large number of ESS can be expected to exhibit instability, chaos, and self-organization.
- These deterministically complex ESS have, by their nature and definition, inherent elements or order and complexity (disorder).
- Deterministic complexity in ESS is therefore a plausible explanation of the simultaneous order and complexity often observed in real landscapes and environmental systems.
- Order and complexity are emergent properties of ESS; that is, one or the other emerges as the dominant mode according to spatial and temporal scales.

4

The General Case

Many (again, not all and not always!) earth surface systems are unstable, potentially chaotic, and self-organizing. I base this assertion on four lines of evidence.

1 *Empirical evidence and case studies* Many studies of specific ESS show that they are unstable, chaotic, self-organizing, or that they exhibit behaviors such as spontaneous pattern formation and increasing divergence over time. The preceding chapter reviewed a number of such studies.

2 *Properties of generalized, canonical models* Many highly generalized models of ESS, which could be expected to represent a wide number of systems, may exhibit deterministic complexity. This includes a broad class of finite difference Verhulst-type models, which may describe phenomena as varied as population dynamics and hillslope evolution, as described later in this chapter. Similar difference equations used to model a broad range of hillslope, soil, and regolith development phenomena may also be chaotic under realistic assumptions (Phillips 1993c,f,h). The fundamental equations of motion for the atmosphere are well-known examples of deterministic chaos (Lorenz 1963, 1965, 1990; Tsonis and Elsner 1989; Palmer 1993; Zeng et al. 1993), as is the phenomenon of turbulent flow, ubiquitous in ESS.

3 *Deterministic uncertainty* The spatial expression of any nonlinear dynamical ESS that is subject to an underlying deterministic constraint must have finite positive K-entropy.

4 *Differentiation and divergence* Many evolving ESS become increasingly differentiated over time (for example, increasing biodiversity in ecosystems, or relief-increasing landscape dissection). An ESS where differentiation is oc-

curring must, at some scale, be unstable, chaotic, and self-organizing.

Empirical evidence and case studies (point 1) have already been treated, and generalized models are discussed below. Points (3) and (4) are formal, general arguments that (assuming you accept them) give weight to the empirical evidence represented by the first two points.

Chaotic Behavior in Generalized Models

Lorenz equations

The two most famous and widely cited examples of mathematical chaos are the logistic map, treated below, and the Lorenz equations. The latter system was devised in Edward Lorenz's (1963) article, a seminal paper in the nonlinear dynamics literature and increasingly important in the atmospheric sciences literature. The Lorenz equations give an approximate description of a horizontal fluid layer (in this case, the atmosphere), heated from below. The fluid at the bottom warms and rises to create convection.

$$dx/dt = -ax + ay \qquad (4.1a)$$

$$dy/dt = -xz + bx - y \qquad (4.1b)$$

$$dz/dt = -xy - cz \qquad (4.1c)$$

x is proportional to the intensity of the convective motion, y is proportional to the horizontal and z to the vertical temperature variation, and a, b, and c are constants. Any point in the system phase space describes some combination of x, y, and z. Despite its simplicity, the Lorenz system is chaotic. The evolution in phase space is completely deterministic, but nonperiodic. Initially similar points diverge exponentially over time, and predictability of the system state diminishes rapidly with each iteration. All trajectories (successive system states for different initial conditions) converge onto the familiar "butterfly wing" attractor (figure 4.1), but any two initially nearby trajectories diverge.

The climatological and meteorological implications of the chaotic behavior of the Lorenz equations have been discussed at length elsewhere (Lorenz 1963, 1990; Tsonis and Elsner 1989; Palmer 1993; Zeng et al. 1993). For our purposes, the most important implication is that there may be a climatic "strange attractor;" that is, the internal dynamics of the climate system are inherently chaotic and lead to complex behaviors independently of any stochastic forcings. As many ESS are driven or strongly influenced by climate, or contain climatic components, this implies unstable, chaotic elements in those ESS. Complex responses, lags, and other

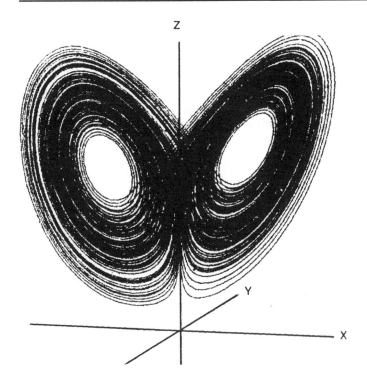

Figure 4.1
The famous "butterfly wing" Lorenz attractor. A numerical solution of the Lorenz equations (equations 4.1) with a = 10, b = 28, and c = 8/3.

environmental controls may complicate ESS response to climate change so that ESS may or may not reflect the rate, magnitude, and variability of climate (Pennington 1986; Bull 1991; Huggett 1991; Christofolletti 1993; Tausch et al. 1993; Kupfer and Cairns 1996). However, chaotic climate change would at least indicate the need to seriously consider the possibility of instability and chaos in a number of ESS.

The chaotic behavior of the Lorenz equations is obvious, and the implications thereof that weather may be inherently unpredictable more than a few days into the future is widely accepted, but the existence of a chaotic attractor at the climatic scale is more controversial (see Bhattacharya 1993). Nicolis and Nicolis (1984), apparently the first to apply Lorenz's ideas to climate, found evidence of a climate attractor, as did Fraedrich (1986, 1987), among others. Lorenz (1990) linked inter-annual climate variability to chaotic atmospheric dynamics. Palmer (1993) extended the Lorenz model to develop a paradigm of the long-range predictability of the atmosphere based on attempts to forecast changes in climate ensembles or regime transitions rather than instantaneous conditions or time-averaged conditions. Pielke and Zeng (1994) showed that the Lorenz model can (but not necessarily that it does) account for long-term variability in climate. Unstable or chaotic dynamics have also been shown to be capable of explaining well-known climatic transitions, such as wet–dry cycles in Africa (Demaree and Nicolis 1990) and ENSO phenomena in the Pacific (Elsner and Tsonis

1993; Chang et al. 1994). We move now to examine some generic models of more direct relevance to earth surface systems.

The logistic map

The Verhulst equation is a simple mathematical depiction of a system where the state of the system at a given time (X) is a function of the state at the previous time increment (X_0), an intrinsic rate of growth (R), and a system attractor (K), generally representing some maximum value:

$$X = RX_0[(X_0 - K)/K] \tag{4.2}$$

The best-known example is the logistic equation or logistic map for population growth,

$$X = RX_0(1 - X_0) \tag{4.3}$$

where X is in this case population size in each generation, R is the per capita rate of population increase when resources are unlimited, and X is scaled to vary between 0 and 1, the latter corresponding to the environmental carrying capacity.

According to the value of R, equation 4.3 may behave in three ways: convergence to a single, stable population level; periodic oscillation between two or more population levels; or deterministic chaos, where the population in succeeding generations is a complex, pseudo-random sequence. R can take values from 0 to 4. For $R < 1$, X converges to 0 (i.e. a population dies out). For $1 < R < 3$, X converges to a single steady-state value equal to $1 - 1/R$ (a fixed-point attractor). Between 3 and approximately 3.57, the behavior is cyclic. There are a series of periodic attractors containing 2, 4, 8, 16, ..., 2^k points, with increasing cycle complexity and length. When $R > 3.57$, the behavior is chaotic and includes cycles of every possible length. Detailed explorations of this class of model are provided by May (1973, 1976) and Schaffer (1985), among others; it is also standard fare in most textbooks on chaos and nonlinear mathematics.

The Verhulst equation and logistic map is "canonical" in the sense that a wide variety of mappings share its essential dynamic properties. These are systems where the value of some factor is a function of the value at the previous time increment and a growth parameter (May 1973, 1976; Schaffer 1985). To quote Allen (1990: 283), "it is a scientific embarrassment that the dynamic complexity of this model and its close relatives went unnoticed until it was revealed in May's classical papers. The implications are very fundamental for science. If such simple models can have complex random-looking dynamics, then how much 'random' behavior is really deterministic nonlinear dynamics in disguise?"

The general behavior of logistic maps and Verhulst-type equations – fixed-point, periodic, or chaotic attractors, depending on parameter values – is apparently insensitive to a number of modifications to the equations. Allen (1990) showed that increasing lags or delays increase the likelihood of chaos. The inclusion of periodic forcings (for example, climate seasonality) and increases in system dimension (i.e. adding more components, such as species in population models) also increases the likelihood of chaos. In general, from the ecological perspective, "the more realism one adds to these models the greater is the chance of chaos" (Allen 1990: 295). Nijkamp and Reggiani (1992) also show explicitly how increasing the number of delays (i.e. the value at time t is a function of more than one preceding value) increases the chaotic parameter space.

Population models are the best-known application of logistic models, but Verhulst-type equations have been applied to several earth surface processes. Afouda (1989) developed a discrete form of equation describing the evolution of the Reynolds number (Re) in fluid flows, a fundamental hydraulic parameter indicating the ratio of driving and resisting forces. This turns out to be a form of the logistic equation:

$$Re_{t+1} = Re_t + a_0 Re_t (1 - Re_t / Re_c) \qquad (4.4)$$

where a_0 is a parameter describing the rate of change of velocity and mean depth and Re_c a critical value of the Reynolds number. Procaccia (1988) shows how the essential dynamics of fluid mechanical systems can be described by a simple one-dimensional transformation of the familiar form

$$x_{n+1} = r x_n (1 - x_n) \qquad (4.5)$$

where x is any realization of system behavior (for example, velocity) and the parameter r encompasses the fluid mechanics involved.

Haigh (1988) applies a Verhulst model to the study of landslide behavior, where the size of landslide features at a given time is a function of the size at the previous time increment and the rate of growth or degradation. Nijkamp and Reggiani (1992) deal with dynamic logit models describing spatial interactions between discrete locations. While their concern is economic geography, the class of models they deal with can apply to any situation where interactions between locations (for instance, water, sediment, or nutrient transfers between sites) can be expressed as probabilities. They obtain a form of the canonical logistic map whereby the quantity of interest is the probability of a particular event occurring in a binary alternative situation (for example, mass is or is not transported) and the parameter describes the attraction of the destination (for example, a function of the magnitude of a flux gradient such as slope or concentration).

I have previously used variations of logistic equations of the general form $x_t = f(x_{t-1})$ both to illustrate the sometimes chaotic

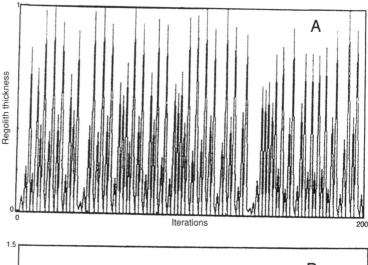

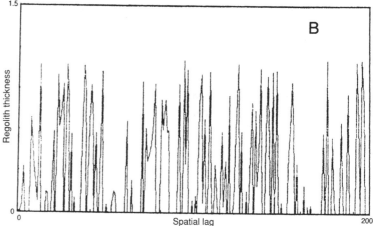

Figure 4.2
Deterministic chaos in the evolution of regolith thickness, modeled as equation 4.6, with $k_1 = 1$. (A) is in the temporal domain, with $k_2 = 0.93$ and $c = 2.95$, and illustrates changes in regolith thickness over time. (B) is in the temporal domain, with $k_2 = 0.99$ and $c = 3$. The graph shows the pattern of regolith cover along a 200-unit spatial gradient after chaotic evolution for 50 iterations at each site. At the outset regolith thickness varied randomly between 0 and 0.001, around a mean of 0.0001 (from Phillips 1995e).

behavior of simple deterministic models and to examine some specific ESS. In particular, the negative exponential parameter may be applicable to the evolution of regolith covers and soils:

$$S_t = S_{t-1} + k_1 e^{-cS(t)} - k_2 \tag{4.6}$$

Here S is the degree of soil development or the thickness of the regolith cover. The constants (k) describe the maximum, climatically controlled rate of progressive pedogenesis or debris production by weathering (k_1) and the rate of regressive pedogenesis or of regolith removal by erosion (k_2). The coefficient c describes the decline in progressive pedogenesis as the soil matures or the decline in weathering at the weathering front as the regolith thickens. The model produces chaotic behavior at realistic parameter values as shown in figure 4.2 (and also, of course, stable, nonchaotic behavior at some parameter values). Applications to regolith thickness are described in Phillips (1993h) and to soil development in Phillips

(1993c). It can also be shown that the model is chaotic in the spatial as well as the temporal domain (Phillips 1993f).

Logistic-type models, then, are applicable to a variety of ESS, and the chaotic properties of this generalized, canonical model strongly suggest that much complexity and apparent randomness in environmental systems and landscapes could be, as Allen (1990) puts it, "deterministic nonlinear dynamics in disguise." However, it is not wise to proceed too far based on this line of reasoning alone, for several reasons:

1 *Continuous (as opposed to discrete) formulations of the models may have analytical, stable, nonchaotic solutions.* This is true of the continuous form of the basic logistic map, and is true of many variants. For example, the continuous form of equation 4.6 ($dS/dt = k_1 e^{-cS(t)} - k_2$) has an analytical solution

$$S_t = 1/c \ln[k_1/k_2 + Ce^{-k_2 t}] \tag{4.7}$$

The generational nature of population change makes a discrete form appropriate for population models. The discontinuous, episodic nature of geomorphic, hydrologic, and pedologic evolution also argues in favor of discrete rather than continuous representations. However, the difference or discrete forms will not always be appropriate.

2 *Logistic-type models are often too simplistic.* While making the models more realistic may increase the likelihood of chaos occurring or enlarge the range of parameter values at which chaos occurs (Allen 1990; Nijkamp and Reggiani 1992), even expanded logistic-type models may be far too limited to provide reasonable representations of many earth surface systems (Malanson et al. 1992).

3 *Realistic modifications may eliminate chaotic behavior.* I re-evaluated the model of equation 4.6 for situations where there is a one-increment lag; that is, regressive pedogenesis or erosion can operate only on soil or debris available at the beginning of a time increment or episode, and is thus dependent on system state at time $t - 1$, while debris produced during increment t is not removed. This simple modification, realistic for many landscape situations, eliminates chaotic behavior at all parameter values (figure 4.3; Phillips 1995e).

Deterministic Uncertainty

Now to the first of the formal, general arguments that many ESS are unstable and potentially chaotic. Any nonlinear dynamical ESS that is subject to an underlying deterministic constraint must have finite positive K-entropy. Because Kolmogorov entropy equals the sum of

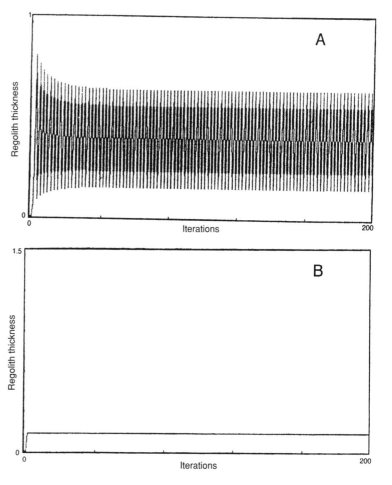

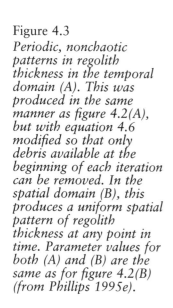

Figure 4.3

Periodic, nonchaotic patterns in regolith thickness in the temporal domain (A). This was produced in the same manner as figure 4.2(A), but with equation 4.6 modified so that only debris available at the beginning of each iteration can be removed. In the spatial domain (B), this produces a uniform spatial pattern of regolith thickness at any point in time. Parameter values for both (A) and (B) are the same as for figure 4.2(B) (from Phillips 1995e).

the positive Lyapunov exponents, and because a positive Lyapunov exponent indicates chaotic behavior, this is indicative of deterministic uncertainty which may be interpreted as chaos. As we shall see, however, the deterministic uncertainty view of chaos is somewhat at variance from that of mainstream chaos theory.

At a given time any ESS will have a particular spatial expression. This may take the form of continuous or discrete representations of variables or components describing or measuring the system; for example, a soil or topographic map, a digital elevation model, or a cellular grid of environmental data. The entropy of the spatial distribution of any variable of an ESS – elevation, pH, soil thickness, hydraulic conductivity, or whatever – can be measured. For an ESS with q components, let's say k_1 is the component of interest and k_i, $i = 2, 3, \ldots, q$ are variables or components which partially determine the value of k_1. The value of any k at a given location is the *state* of the system or component. For example, ks may represent landform elements, soil types, vegetation communities, geological formations, permeability or elevation classes, and soon. The $k_{i>1}$ partly determine k_1 by constraining the state k_1 may be in. Then entropy (H) is

$$H(k_1, k_2, \ldots, k_q) = \sum_{i=1}^{q} H(k_i) - \sum_{i=1}^{q} [H(k_i) - H_{i+1}(k_i)] - H_u(k) \qquad (4.8)$$

Entropy is maximized where any value or state of k can exist at any location, and minimized where only one value or state of k is possible. Equation 4.8 shows that the total entropy of the joint spatial distribution of all q components can be decomposed into that associated with the known, measured constraints $[\Sigma H(k_i) - H_{i+1}(k_i)]$ and that associated with unknown or unmeasured constraints or with white noise $(H_u(k))$.

One example would be a data matrix or collection of data points where the variable of interest (k_1) is soil type, and the constraining variables $(k_{i>1})$ represent parent material, slope elements, and drainage class. For another example, the state of k_1 might be represented by a particular vegetation community, and the constraining variables might be factors such as soil type, drainage class, slope, microclimate, elevation, or pH.

Now suppose that $H_u(k)$ includes the effects of a known, but unmeasurable, constraint. These situations are not uncommon, as illustrated by the phenomenon of biotic influences on soil morphology. Variations in soil morphology, such as depth to the B-horizon, are sometimes related to the effects of trees, for example (Zinke 1962; Gersper and Holowaychuck 1970; Phillips et al. 1994), but it is impossible to discern exactly where trees might or might not have existed over the history of soil development. Likewise, faunal effects may be critical in the formation of texture-contrast soil profiles (Paton et al. 1995), but one cannot measure the extent of faunalturbation at a given location over pedogenetic time scales. Another general example of known but unmeasurable constraints is geological controls over geomorphic evolution. The underlying pre-Holocene topography and lithology exerts definite controls over Holocene geomorphic evolution of some coastal landforms (cf. Kraft 1971; Riggs et al. 1995), for instance, but it is often impractical to collect detailed information on subsurface features over large areas. Twidale (1993) shows how lithologic variations in weathering susceptibility produce landscapes dominated by etching, but where the evolution process itself obscures the initial lithologic controls. Though the underlying geology may be known to constrain surface features, the former may be largely unmeasured.

Let's define ξ as the extent to which the spatial distribution is constrained by the known but unmeasurable underlying factor. There are two extreme cases with respect to $H_u(k)$. At one, all outcomes are equally possible, there are no known underlying constraints, and unless or until some are identified $H_u(k)$ may be considered to be random, white noise effects. In this case $\xi = 0$. At the other extreme, underlying constraints allow only one possible outcome, $H_u(k)$ is completely deterministic, and $\xi = 1$. The extent to which the number of possible spatial arrangements of k has been reduced is given by $1/\xi$. $H_u(k)$ can be decomposed into its random and deterministic components:

$$H_u(k) = H_{\text{ran}}(k) + H_c(k) = H_{\text{ran}}(k) + \ln \xi \qquad (4.9)$$

Thus the deterministic uncertainty is measured by

$$H_c(k) = -\ln \xi \qquad (4.10)$$

If the equations and the argument seem familiar, it is because they are a generalization from chapter 2, where for simplicity's sake we assumed a hierarchical chain of constraint. If the degree of constraint can be estimated – for example, if you can estimate that the effect of trees reduces the variation in depth-to-B-horizon classes by 50 percent so that $\xi = 0.5$ – then you've got a quantitative estimate of deterministic uncertainty. For our purposes now, however, it is sufficient to show that the very existence of an underlying deterministic constraint indicates the presence of a form of chaos.

Note that this differs substantially from the classic conception of chaos as irreducible complexity which is immune to reductionist efforts at reducing uncertainty. In this case, it is at least theoretically possible that improved measurement technologies or unusual sampling opportunities would allow full indentification and measurement of the underlying constraint. However, this deterministic uncertainty is consistent with basic ideas of dynamic instability in that minor variations and perturbations that are no longer apparent, or perhaps no longer even in existence, persist and grow over time. This situation also shares with classic conceptions of chaos the idea that a degree of uncertainty and complexity is inherent in the system structure and has a deterministic source, rather than deriving from stochastic forcings.

Differentiation and Divergence

Many ESS become increasingly differentiated over time. That is, an initially more-or-less uniform landscape becomes ever more differentiated into distinct spatial elements. A good intuitive example is relief-increasing topographic evolution. From whatever initial topographic state, dissection produces a landscape increasingly differentiated into more distinct elements based on elevation, slope, or morphology. Other examples include channel network development, whereby initially unchanneled terrain is progressively differentiated into channel and hillslope units, and many cases of ecological succession, where early communities with a single dominant species may be gradually replaced by more diverse communities with several co-dominant species.

Let us formalize a generalized ESS by designating the component of interest (elevation, soil thickness, species richness, or whatever) as x, and letting the subscripts i and j denote any two locations within the ESS. If differentiation is occurring, then, on average

$$\frac{d(|x_i - x_j|)}{dt} > 0 \tag{4.11}$$

In chapter 2 we noted that Rosenstein et al. (1993) showed that randomly chosen pairs of initial conditions diverge exponentially (on average) in a chaotic system at a rate given by the largest Lyapunov exponent λ_1. Thus, where $d(t)$ is the average divergence of randomly selected pairs at time t and C is a constant that normalizes the initial separation, equation 2.13 gave us

$$\lambda_1 = \ln d(t) - \ln C$$

Let us suppose we start with a high-relief landscape where the mean initial elevation difference between randomly selected locations is 100 m. After a few millennia of downwasting, let's say relief has decreased as hilltops erode and valleys aggrade, and the mean elevation difference is 50 m. Then $\lambda_1 = -0.22$, and we have a stable, non-chaotic system. But many ESS don't work that way. Bishop and others (1985), for example, found that in the east Australian highlands both the modern and early Miocene evolution of rivers indicated some increase in relief, so that the average elevation difference must increase over time. Twidale (1991) shows that relief may increase over time in a variety of tectonic settings, and over time scales as long as 60–100 million years. The mean separation between locations in drylands in terms of soil nutrients has also been shown to increase over time, particularly during soil degradation (Schlesinger et al. 1996). In parabolic dune systems of eastern Australia, as the age of dunes and associated soils increases, the variability of the thickness of E- and B-horizons increases, and thus the mean separation of individual locations (Thompson 1983; Thompson and Bowman 1984). Numerous other examples of divergence over time are given in chapter 3.

Now let us suppose that we have an initial low-relief landscape, where the mean elevation difference is only 10 m, and after a period of dissection the mean elevation difference has increased to 20 m. In this case $\lambda_1 = 0.69$, and the system is unstable and chaotic. Not all ESS are characterized by increasing divergence over time. But many clearly are, and this is clear and intuitively simple evidence that instability and deterministic complexity must be quite common.

Summary and Synthesis

Instability, deterministic uncertainty, and chaos are ubiquitous in earth surface systems. While stable systems are also common, they

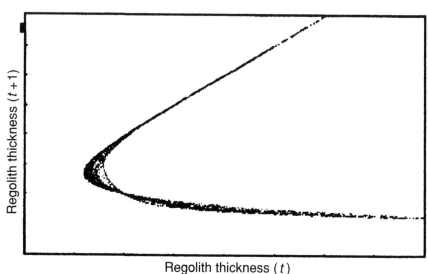

Regolith thickness (t)

Figure 4.4

Poincairé map for equation 4.6 with chaotic parameter values ($k_2 = 1$; c varying from 2 to 4. The map is produced by plotting values of regolith thickness at time t with those at iteration t + 1. In this figure there were 200 iterations for each 0.1 increment of c, but for economy only every sixth pixel was plotted. The Poincairé map illustrates how a chaotic system none the less occupies a restricted portion of phase space, unlike a random white noise system, where any combination of values is possible.

are not necessarily the norm; there may not be a norm in this regard. Numerous real-world earth surface systems and plausible models thereof show evidence of instability, chaos, and self-organization. This circumstantial, inductive evidence mounts with every new batch of scientific journals. The generality (but not universality, since results of negative searches for chaos are also flowing forth) of these findings is bolstered by the fact that chaotic dynamics appears to be common in meteorological and climatological phenomena, one of the fundamental forcings of ESS. The generality of chaos is also suggested by the chaotic behavior of canonical models such as the logistic difference equation. To take the argument a step farther, formal demonstrations have shown that, under certain circumstances, a wide variety of ESS must have unstable, chaotic elements. This applies to ESS with known but unmeasurable underlying constraints, and to those ESS which exhibit spatial divergence over time.

In chapter 3 the simultaneous presence of both order and disorder in many real landscapes and environmental systems was documented. The common presence of deterministic complexity in ESS is a plausible explanation of this phenomenon. Chaotic dynamics, complicated, irregular, and apparently random as they may be, are none the less governed by an attractor in the state space defined by all the variables or components of the system. Unlike a truly random system, where the system state might occupy any point of the state space with equal probability, the chaotic attractor dictates that the chaotic system occupies only a restricted portion of the space, and further occupies some portions of that restricted area with greater probability than others (figure 4.4). Thus, as scale is broadened the irregular chaotic details become less important and apparent, and the broader-scale structure with well-defined limits

dominates the view: order emerges from chaos. Order is an emergent property of the unstable, chaotic system. In the other direction,
as scale is narrowed, only a single realization of the system, or a few
consecutive realizations, come into view. These are governed by
deterministic dynamics and, while the long-term evolution may be
inherently unpredictable, the state of the system for the next increment or two is perfectly predictable. Once again, regularity and
order emerge from chaos as scale changes.

The fundamental implication is that while instability and chaos
are not necessary to give rise to simultaneous order and complexity,
they are sufficient to do so. Further, an unstable/chaotic system
dictates that elements of both regularity and irregularity must exist
in an ESS. Finally, unstable dynamical systems also ensure that both
regular/orderly and irregular/complex patterns must emerge as the
spatial and/or temporal scale or resolution is changed. Both order
and disorder are thus emergent properties, with corresponding
implications for our ability to understand ESS by building up from
process mechanics or splitting down from broad-scale structures.

5

Solving Problems

Thus far, we have learned something of the nonlinear dynamics perspective of the study of earth surface systems, examined evidence of both order and disorder in ESS, and made a case that a good many ESS are unstable, potentially chaotic, and self-organizing. From that, we have advanced the arguments that unstable nonlinear dynamics explains the ubiquity of simultaneous order and disorder in ESS, and that (in)stability, (dis)order and related phenomena are emergent properties of ESS which appear or disappear as scale or resolution is changed.

Before moving on, we will in this chapter take the opportunity to examine a few ESS problems in a bit more detail to illustrate some points and bolster some assertions made earlier. The problems chosen are biased toward my own previous work and topics within my realm of experience. None the less, the menu of problems is large enough that we will be able to examine various issues illustrating some prototype problems in ESS. These may be canonical in the sense that their general characteristics represent a broader class of problems in geomorphology, pedology, hydrology, and environmental science.

The four case studies will, in turn, use an NDS approach to:

- Examine a classic example of mutual adjustments in ESS (hydraulic geometry).
- Attempt to shed light on a problem where a plethora of models and explanations has led to confusing and sometimes contradictory results (channel network evolution).
- Address a classic question in the earth sciences (evolution of topographic relief).
- Confront the complex interactions of the atmo-, litho-, hydro-, and biospheres in the context of a critical environmental problem (desertification).

Hydraulic Geometry

Hydraulic geometry involves the interactions between discharges through a stream channel and the channel itself, and the manner in which changes in discharge are accommodated at a given cross-section. Flows help shape the channels they flow through. Channels help determine the character of the flows through them. Hydraulic geometry, then, represents an examplar of mutual adjustments in earth surface systems. The study of hydraulic geometry is important in its own right due to its significance for making process interpretations based on channel morphology in historical geomorphology and paleohydrology, and its utility in predicting channel response to imposed flows in process geomorphology, hydrology, hydraulic engineering, and water resource management.

At-a-station hydraulic geometry concerns the variations in flow width, depth, and velocity associated with variations in discharge. If flows cannot or do not modify the channel, and external factors do not alter the cross-section, Ferguson (1986) shows that the problem is a straightforward one involving hydraulics of flow resistance and geometry of the channel cross-section. But since flows often can alter the channel itself, and external factors can influence both geometry and hydraulics at any point along a stream, these external perturbations and mutual adjustments make the problem more complex. Hey (1978, 1979) was one of the first to approach it from a systems perspective, noting that at-a-station hydraulic geometry can involve trying to solve the simultaneous, mutual adjustments of as many as nine variables.

Solving the hydraulic geometry problem is difficult because of indeterminacy (more variables than equations to solve them; Hey 1978; Ferguson 1986) and the debatable nature of the hypotheses and assumptions used to derive solutions (Griffiths 1984; Ferguson 1986; Lamberti 1988; Phillips 1991d). Observed relationships are quite variable, and existing hydraulic geometry equations are physically questionable (Richards 1982; Griffiths 1984; Knighton 1984; Ferguson 1986; Lamberti 1988).

The analysis here, based on Phillips (1990d), does not seek to model stable or most probable channel configurations for a given discharge. Rather, we examine the broader question of how the channel system responds to changes in system components.

At-a-station hydraulic geometry

Leopold and Maddock (1953) developed a well-known set of empirical power functions relating width, depth, velocity, and other variables to power functions of discharge. The three most important are

$$w = aQ^b \tag{5.1}$$

$$d = cQ^f \qquad (5.2)$$

$$v = kQ^m \qquad (5.3)$$

where w is width of the water surface, d is mean depth, v is mean velocity, Q is discharge, and a, c, k, b, f, and m are coefficients. The continuity equation dicates that the product of a, c, and k and the sum of b, f, and m are both 1. These and similar relationships for other hydraulic variables fit data well in some instances, but there is no theoretical reason why the equations should be in the form of power functions (Ferguson 1986; Lamberti 1988).

Hydraulic geometry theory has often been based on "extremal" hypotheses to identify constraints or boundary conditions to estimate the coefficients in the power functions. The minimum variance theory (Langbein 1964) determines the most probable values for the hydraulic geometry exponents by minimizing their collective variability (see the review by Williams 1978). Other extremal hypotheses include minimum stream power, minimum unit stream power, minimum energy dissipation, maximum sediment transport, maximum friction factor, and maximum slope. Extremal hypotheses are also employed in downstream hydraulic geometry (or river regime). Four independent relationships are necessary to link the variables for a determinate solution, but only three are readily available: flow continuity, flow resistance, and sediment transport capacity equations. The fourth is controversial, and extremal hypotheses have been used to obtain closure. Review, discussion, and both positive and negative critiques of extremal hypotheses for both downstream and at-a-station hydraulic geometry are given elsewhere (Davies and Sutherland 1983; Ferguson 1986; Lamberti 1988, 1992; Phillips 1990d, 1991d; Yang 1992).

Extremal hypotheses are essentially variational approaches, whereas vectorial approaches are based on the force, momentum, and resistance in the cross-section (Yang 1992). While vectorial approaches have been quite successful in problem-solving and have provided important insights into hydraulic geometry and flow–channel interactions, these approaches are often difficult to generalize beyond a specific cross-section because a lack of full specification in vectorial models leads to a reliance on assumptions, empirical data, and model parameters. Vectorial approaches are also inapplicable when constraints are not easily replaced by forces. Thus, following Yang (1992), variational methods are preferred for attempts to develop generalizations about hydraulic geometry relationships and behaviors.

The fundamental problem for hydraulic geometry is that there are more unknowns than equations to describe them, precluding a unique solution (Knighton 1984: 88; Ferguson 1986), at least until satisfactory physical relationships are discovered (Hey 1978, 1979). An NDS approach causes us to focus on the question of how the system responds to changes and disturbances, rather than the specific responses of system components, allowing the determinacy problem to be circumvented.

Theory

Discharge, width, depth, and velocity are related by the mass continuity equation

$$Q = wdv \qquad (5.4)$$

where Q is discharge (m^3/s), w is water surface width (m), d is mean water depth (m), and v is mean velocity (m/s). A change in any variable must be accommodated by an appropriate combination of the other three.

Changes at a cross-section may be indirect, and may influence more than one variable (plate 5.1). Discharge variations may simultaneously affect w, d, and v, for example. Channel debris may influence velocity indirectly via roughness, channel shape, and cross-sectional area. The other variables (in addition to Q, w, d, and v) necessary to describe the system are flow resistance and slope. The addition of these variables, along with the constant of the specific gravity of water (ρg), allows a fully specified, completely determined system: that is, every variable can be described as some function of the other variables. Using s to denote energy gradient (commonly approximated as water surface slope), the Darcy–Weisbach friction factor (f) to describe flow resistance, and the Darcy-Weisbach flow resistance equation:

$$Q = wd \left(\rho g R s / f \right)^{0.5} \qquad (5.5)$$

Plate 5.1
An urban river channel: the Fiume Adige, Verona, Italy. River channels and the flows they carry mutually adjust. There are multiple ways in which flow hydraulics and the channel itself – even with fixed banks – may respond to changes in discharge, including velocity, roughness, width, depth, and slope.

where hydraulic radius $R = wd/(2d + w)$. Equation 5.5 can be rewritten to solve for any variable on the right of the equality.

If all other variables can be held constant, it is quite simple to predict channel response to change in any one variable. But because the variables are mutually adjusting, this is unrealistic. Equation systems based on equation 5.5 can be solved iteratively for a specific case if the relative rates of change of the variables are known. However, such an iterative solution would not allow generalizations because the relative rates of change of variables in different stream systems vary substantially, in response to factors such as stream competence or the availability of debris.

Two arguments can be invoked to collapse the hydraulic geometry system into a compact model including velocity, hydraulic radius, slope, and the friction factor (Phillips 1990d). One can start by identifying the minimum number of variables which subsume the variables Q, w, d, v, f, and s. The argument is that the omitted variables (Q, w, d) are subsumed in the included variables. Width and depth are subsumed by hydraulic radius, as R is a unique function of w and d. An infinite number of combinations of w and d can produce the same R, but the hydraulic influences of w and d are exerted via hydraulic radius. Because Q is a unique function of w, d, and v, discharge is a unique linear function of the product Rv. Because Q, w, and d are subsumed in the variables R and v, the former group may be omitted from the analysis, leaving R, v, f, and s.

The second justification is based on rules for system aggregation, applied to the nine-variable hydraulic geometry equation system developed by Hey (1978). The latter represents the nine degrees of freedom a stream has to respond to changes in discharge or other controlling factors: R, s, and v, wetted perimeter, maximum depth, channel sinuousity, meander arc length, dune height, and dune wavelength. The system is intractable because there are not nine equations for the nine variables (Hey's expressions are basically formal statements) and because a nine-variable system is too large for effective stability analysis. Slingerland (1981) let R represent wetted perimeter and maximum depth, f represent the bedform factors, and assumed a straight reach to eliminate the channel planform variables. He also added a variable (bedload transport rate).

The rules and criteria for lumping variables for analysis of system models (see Puccia and Levins 1985: 79–81) can be used to aggregate or eliminate variables and reduce the system. The rules are such that using them will not influence the qualitative outcome of the analysis. One rule is that a subsystem of variables linked by one variable to the rest of the system may be replaced by a single variable. This can be used to aggregate wetted perimeter, maximum depth, and hydraulic radius into a single variable represented by R. The hydraulic influences of depth and wetted perimeter on the

Table 5.1 Interaction matrix derived from equations 5.6a–d. Each entry a_{ij} represents the positive, negative, or absence of influence of the row variable on the column variable

	v	R	s	f
v	0	a_{12}	a_{13}	$-a_{14}$
s	a_{21}	$-a_{22}$	$-a_{23}$	a_{24}
R	a_{31}	$-a_{32}$	0	a_{34}
f	$-a_{41}$	a_{42}	a_{43}	0

v, velocity; R, hydraulic radius; s, slope; f, friction factor.

other variables are expressed through R, as reference to any standard flow equation attests. The bedform and planform factors can similarly be lumped into the resistance factor, f (which is also influenced by bank roughness, cross-sectional shape, and obstructions such as vegetation). Influences of the variables listed above on channel hydraulics can be assumed to be exerted entirely via flow resistance, allowing substitution of f without changing the qualitative outcome of the analysis.

Hence nine variables are reduced to four. These can be linked by rearranging the Darcy–Weisbach equation, with each equation written so that the positive or negative relationships are easily seen from the exponents.

$$v = R^{0.5} s^{0.5} f^{-0.5} (\rho g)^{0.5} \tag{5.6a}$$

$$s = v^2 R^{-1} f^1 (\rho g)^{-1} \tag{5.6b}$$

$$R = v^2 s^{-1} f^1 (\rho g)^{-1} \tag{5.6c}$$

$$f = v^{-2} R^1 s^1 (\rho g)^1 \tag{5.6d}$$

This equation system is translated into an interaction matrix A (table 5.1). The magnitude of the influence can be determined from the values of the exponents, and signs of the entries are determined from equations 5.6a–d, with the constant (ρg) omitted. One entry $(-a_{22})$ is not derived from the equations, and represents negative self-damping feedback of energy grade slope. This self-regulating loop was included because at least one self-damping feedback relationship is a necessary and sufficient condition for stability of an interaction matrix. The self-regulating link can be physically justified by the self-adjusting tendencies of water surface profiles which are implicit in the Saint-Venant equations. As we know from chapter 2, equations 5.6a–d or table 5.1 allow us to determine the qualitative dynamic properties of the system.

Results

The characteristic equation is

$$\lambda^4 + a_{22}\lambda^3 + (a_{12}a_{21} - a_{23}a_{32} - a_{43}a_{34})\lambda^2 + (-a_{12}a_{31}a_{23} + a_{13}a_{21}a_{32} - a_{13}a_{31}a_{22} - a_{22}a_{43}a_{34} + a_{23}a_{42}a_{34} + a_{24}a_{32}a_{43})\lambda$$
$$+ \det A = 0 \tag{5.7}$$

where each a_{ij} is an entry in the interaction matrix.

F_2 must always be positive, violating the Routh–Hurwitz criteria. This can be demonstrated by writing the coefficient with the signs of each a_{ij} indicated:

$$F_2 = a_{12}a_{21} + (-a_{23})(-a_{32}) + a_{34}a_{43} \tag{5.8}$$

Not all eigenvalues have negative real parts, there must be a positive λ, and the system is unstable and potentially chaotic.

Implications and examples

Unstable hydraulic geometry implies that not even the qualitative relationships among the hydraulic variables, not to mention the quantitative ones, are likely to persist in the face of changes in imposed flows or other perturbations to the cross-section. How is this manifested in fluvial behavior? First, the instability suggests multiple modes of adjustment (MMA), where a mode of adjustment is a specific combination of increases, decreases, or relative constancy of hydraulic variables in response to imposed changes. For example, an increase in discharge accommodated by an increase in mean depth and a lower friction factor, while s and v remain constant, would constitute one of many specific modes of adjustment. Second, unstable chaotic behavior implies opposite-from-expected behavior. For example, the Darcy–Weisbach equation would suggest that if Q decreases, then v, R, and s should decrease and f should increase. However, if the falling Q were accompanied by an *increase* in velocity, offset by changes in one or more other variables, the decline in v would represent opposite-from-expected behavior.

An example is a series of detailed flow measurements in the natural channel of the Bogue Phalia river near Heads, Mississippi, originally presented by Fasken (1963), and reworked to investigate the stability of hydraulic geometry by Phillips (1990d). Figure 5.1 shows discharge, velocity, surface width, hydraulic radius, slope, and roughness (f) for nine measured flows. There is no consistent mode of adjustment. Consider the eight transitions from one discharge to another, from left to right. There are five different modes of adjustment. In three cases, all variables respond exactly as expected to increasing Q (increasing v, w, R, s; decreasing f). In one case, v, s, and f respond to increasing Q by decreasing, but the

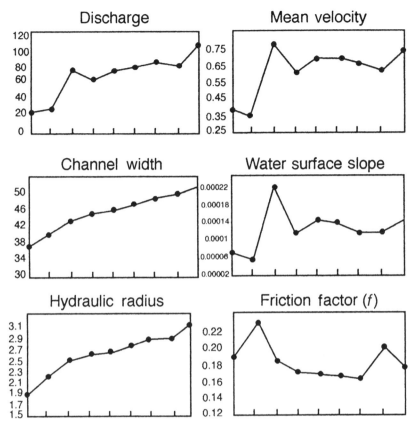

Figure 5.1
Variation of discharge and hydraulic variables over nine flow events in the Bogue Phalia river, Mississippi from the data of Fasken (1963).

higher flow is accommodated by increasing w and R. In another case, higher discharge is accompanied by constant v, lower s and f, and higher w and R. In two cases, a decrease in discharge is associated with (as would be expected) lower v and s, which offset a decline in f and increases in w and R. The fifth mode of adjustment is a response to decreasing Q via decreasing v and increasing f to offset increases in w, R, and s.

The Bogue Phalia data provide an excellent illustration of unstable at-a-station hydraulic geometry with multiple modes of adjustment and opposite-from-expected behavior. MMA and opposite-from-expected change are also shown in data from four alluvial rivers, and from overland flow experiments, presented by Phillips (1991d, 1992c). MMA can also be inferred from the data of Simon and Thorne (1996) for the North Fork Toutle river, Washington, Ergenzinger (1987) for the Butramo river, Italy (figure 5.2), and Brush (1961) for Pennsylvania streams.

MMA are also inherent in extremal hypotheses. The various extrema are equivalent in terms of their implications as to how the basic hydraulic variables will respond to changes in discharge. There are infinite combinations of specific flow, width, depth, velocity, and slope which can accommodate a given extremum, and

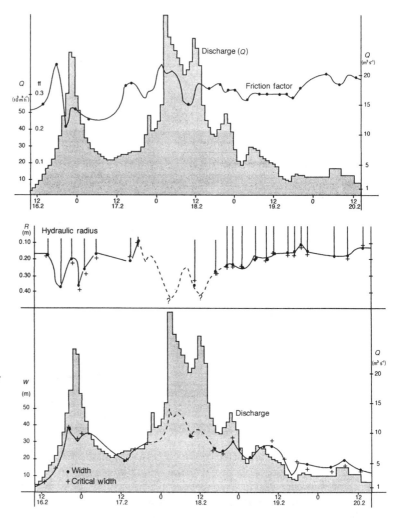

Figure 5.2
*Discharge and hydraulic
variables from
Ergenzinger's (1987) study
of the Butramo river, Italy,
illustrate multiple modes of
adjustment and opposite-
from-expected behavior. If
all variables behaved as
predicted by hydraulics,
hydraulic radius and width
would always change in
the same, and the friction
factor in the opposite,
direction as discharge.*

no basic principles to exclude opposite-from-expected responses in
individual variables (Phillips 1991d).

More generally, dynamical instability – whereby the details are
often complex and unpredictable, but with broader-scale deterministic order and probabilistic regularity – is indicated by the fact that
only average hydraulic geometry exponents can be reliably predicted, with significant scatter around the trend lines, but the averages are often predicted very well (Williams 1978; Miller 1991;
Ridenour and Giardino 1991). Natural river cross-sections exhibit
more complex behavior than conventional hydraulic geometry
suggests (Richards 1982: 152; Ferguson 1986: 3–5). This observed
variability is entirely consistent with an unstable system, which
would not be expected to maintain a constant relationship between
hydraulic and geometric variables.

Channel Network Evolution

There is a remarkable regularity in the topology of fluvial channel networks. They differ in detail, but both the topological statistics and the general visual impression conveyed are remarkably consistent across a range of spatial scales and environments. This has led to considerable efforts to analyze and model networks in hopes of discovering some sort of underlying truth about network evolution.

Network and basin evolution models

Some models of channel network and drainage basin evolution currently in vogue rely on "least work" principles, whereby the network or basin evolves so as to minimize work or some related quantity or maximize thermodynamic entropy. These ideas revive and revise theories and models prevalent in the 1960s and early 1970s. Leopold and Langbein (1962) maintained that river systems evolve to a "most probable" state associated with maximum entropy, and produced simulated networks with realistic statistical properties based on a random walk. Scheidegger (1967) and Howard (1971) also produced random-branching network models. Woldenberg (1969) argued that the number of stream orders and number of basins per order balances opposing tendencies for minimum overland flow in small basins and maximum work savings in large ones. He was able to make reasonable predictions for 17 real basins based on this model, and showed that thermodynamic entropy was near the maximum possible. Kirkby (1971) was apparently the first to explicitly propose that fluvial system evolution is governed by a principle of minimizing energy dissipation rates. Simon (1992) conducted one of the few field studies on energy minimization, documenting energy-minimization adjustments in a mountain fluvial system disturbed by a volcanic eruption and a coastal plain alluvial stream disrupted by channelization. Note, however, that there are a multitude of adjustments or combinations thereof which can achieve given minima (Phillips 1991d; Troutman and Karlinger 1994), as shown in subsequent work by Simon and Thorne (1996).

Ever more sophisticated probabilistic models continued to appear in the literature (see Abrahams 1984 for a review of work through the early 1980s). Van Pelt and others (1989) proposed that the topology of stream networks (and, significantly, other binary branching patterns) can be regarded as a set of partitions of magnitudes at junctions. Their models based on branching probabilities gave good fits to 13 of 14 data sets from the literature. The model of Troutman and Karlinger (1992, 1994) combined elements of random topology and spatial simulation models, generating networks on a spatial grid, but assigning variable probabilities to possible networks. They found that a Gibbs distribution, with a parameter varying with slope and drainage area, was an improvement over a strictly random model, where channels are equally probable in any grid cell.

In the 1980s, several observers noted that channel networks have non-integer fractal dimensions, and that these are often close to 2, the expected value for a space-filling network. This resulted in a number of papers which fitted fractal statistics to maps or digital elevation models of real networks, produced fractal simulations of networks and basin topography in a variety of ways, or argued that networks evolve to completely drain their catchment, so that the topological fractal dimension approaches 2 (Seiler 1986; Hjelmfelt 1988; Tarboton et al. 1988, 1989; Gupta and Waymire 1989; La Barbera and Rosso 1989; Marani et al. 1991; Rosso et al. 1991; Ijjasz-Vasquez et al. 1992; Karlinger and Troutman 1992; Beer and Borgas 1993; Goodchild and Klinkenberg 1993; Nikora 1994).

Closely related are network simulation models based on various principles of self-organization and self-organized criticality. Some, such as Stark's (1991) self-avoiding percolation model and Moglen and Bras's (1995) model based on horizontal and vertical variations in erodibility, are linked to some controlling geomorphic or hydrologic factor (in this case variable substrate erodibility). Others are simply algorithms that produce plausible channel networks, but with limited hydrologic or geomorphic realism (Meakin 1991; Kramer and Marder 1992; Takayasu and Inaoka 1992; Masek and Turcotte 1993; Sun and Meakin 1994; Sun et al. 1994, 1995).

Subsequently, the fractal (or multifractal) structures of stream networks were linked to least-work principles, just as the random models were earlier. Rodriguez-Iturbe et al. (1992) and Rinaldo et al. (1992) argued that fractal structures arise as a consequence of energy dissipation, and La Barbera and Roth (1994) related fractal dimensions of networks to the cumulative probability distributions for stream order, length, area, and energy dissipation per unit of channel. Other papers explicitly linked the obvious self-organized, fractal structures of channel networks and drainage basins with principles of optimality in the venerable terms of energy dissipation. Optimal channel networks can be obtained by minimizing local and global rates of energy expenditure, and the optimal channel networks are spatial models of self-organized criticality, with fractal structures arising as a joint consequence of optimality and randomness (Howard 1990; Rinaldo et al. 1993). Minimum energy expenditures and self-organization were linked to the distinction between threshold-independent hillslope processes and threshold-dependent fluvial processes by Rigon et al. (1994), who suggested (in agreement with Chase 1992) that diffusive processes tend to reduce fractal dimensions toward more realistic values.

Other successful models (cf. Kirkby 1971) have been based on the principle of instabilities in overland flow. Smith and Bretherton (1972) produced a model whereby instabilities leading to channel initiation occur when advective processes dominate diffusive processes. Small disturbances grow rapidly to form channels.

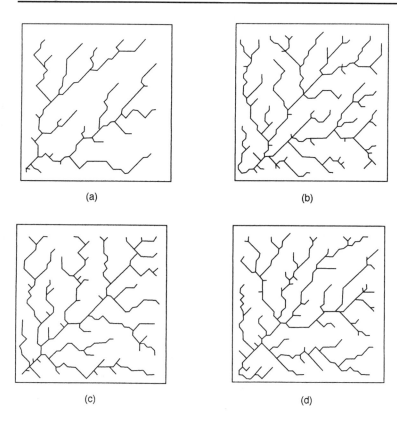

Figure 5.3
Four networks simulated by the process-based model of Willgoose et al. (1991b). Small random variations in initial elevation differences result in the evolution of different networks.

Loewenherz (1991) revisited the Smith–Bretherton model with a nonlinear treatment, and solved the problem of unstable growth of infinitesimally short wavelength rills by introducing a transport function to account for microscale mass transfers. This has the effect of damping very short wavelength rill spacings, and simulating the most rapid unstable growth rates at more realistic moderate wavelengths. Dunne and Aubry's (1986) field experiments confirmed the general model of sheetflow instability leading to separation into rills, and rill incision when advective wash processes dominate over diffusive rainsplash. Tarboton and others (1992) used a version of the Smith–Bretherton model to show that drainage density in a network is related to a transition between the spatial scales where unstable channel-forming processes lead to stable diffusive mass fluxes.

Another general class of models is more physically based, relying on mass transport functions and thresholds of channel initiation. The drainage basin evolution model of Willgoose and others (1991a–c) is based on differentiating between hillslope and channel processes, with channels forming where a channel initiation threshold dependent on flow, slope, and resistance is exceeded (figure 5.3). The model produces realistic-looking basins with appropriate statistics (Willgoose 1994). A division of the landscape into areas dominated by stable diffusive areas and those prone to channel instability due to runoff is the basis of the model devised by Dietrich et al. (1992), who link erosion thresholds to surface morphology in a model of the initiation and growth of channels and headwater valleys. Howard's (1994) basin evolution model goes farther still, distinguishing between creep and more rapid forms of diffusive transport and both detachment- and transport-limited fluvial erosion and transport. The simulated valley systems were found to be similar to those produced by models of "optimal" basins (cf. Rinaldo et al. 1993; Rigon et al. 1994; Sun et al. 1994).

Model equifinality

The fact that more or less realistic stream channel networks can be produced by models with such widely varying assumptions, by so many different (sometimes fundamentally different) algorithms, and purporting to illustrate several different underlying principles, suggests at least two non-exclusive possibilities:

1 This is a case of "model equifinality" (Beven 1996), whereby a wide variety of models produces the same general result, such that the channel network is independent of the rules under which it is modeled.
2 The varied underlying explanations implied in network models are specific cases of some broader general unifying principle.

The network modeling exercises have not gone unquestioned, of course. Abrahams's (1984) review concluded that random topology models were a potentially helpful reference abstraction, whose utility essentially depends on the preferences of the researcher, but that networks are not fundamentally random. Phillips (1993g) reached similar conclusions about fractal models. More fundamentally, Kirchner (1993) showed that "Horton's laws," relating various topological properties of channel networks, are a statistical inevitability of any branching network, given the nature of the stream ordering system. This is problematic, as agreement with Horton's laws is often used to test the realism of network simulations, and Horton ratios are linked to and sometimes used to derive fractal dimensions. Essentially, then, random, fractal, and self-organizing models of channel networks are untestable because they cannot produce anything other than networks whose topology closely resembles that of many real networks (Kirchner 1993, 1994). Claps and Oliveto (1996), for example, show that the fractal properties of river networks are strikingly similar to those of artificially generated deterministic "fractal trees." Liu (1992) undertakes a similar exercise by comparing stream networks with loopless random aggregate trees, with similar results. This may tell us something about the general structure of branching flow networks, but sheds little light on the processes in, or evolution of, fluvial drainage networks.

The fractal and multifractal structures of channels and basins have been linked to physical principles of fluvial system development. However, several authors have pointed out that fractal and multifractal measures are morphometric indices, albeit quite useful and sophisticated indices, that have no direct link to geomorphic or hydrologic processes or environmental controls (Nikora 1991; Phillips 1993g; Howard 1994). While fractal measures can be interpreted as indicative of a space-filling drainage network, or of "optimal" distributions of energy dissipation, they can also be interpreted as simply a handy index of topological complexity or as an index of the extent of geological constraints (Nikora 1991; Phillips 1993g; Nikora et al. 1996). Fractal analyses based, as many are, on extraction of networks from digital elevation models are sensitive to arbitrary selections of the threshold or minimum area necessary to support a channel (Helmlinger et al. 1993). Because any number of real or imagined processes or controls can produce a fractal or multifractal network, the fractal measures tell us no more about network evolution than any other morphometric measure (Kirchner 1994).

Self-organizing models of network and landscape evolution were strongly criticized by Sapozhnikov and Foufoula-Georgiou (1996). They maintain that the final states of the networks and landscape models are not actually critical, attractor states because they evolve to a state where perturbations do not change the configuration of the system. Sapozhnikov and Foufoula-Georgiou also argue that optimality principles of energy dissipation do not follow from the self-organizing fractal models, and that hypothesis tests using model simulations are unreliable. Troutman and Karlinger (1994), in the course of describing their two-parameter Gibbsian probability model of channel networks, also question the assumptions of optimal channel networks. They note that minimum energy is a non-unique configuration, and that energy minimization would not be a valid assumption for all networks in any case.

The criticisms of network modeling, along with the fact that plausible networks can be produced from models with little or no geomorphic or hydrologic basis, suggests that model equifinality clearly exists in this field (Beven 1996). Is there also some broader unifying principle, associated with deterministic complexity?

Entropy and chaos in network evolution

The entropy of a channel network is closely related to network structure and morphometric measures thereof. Elevation (E) is a logical state variable of a network (Gupta and Waymire 1989; Tarboton et al. 1989; Fiorentino et al. 1992). Fiorentino and others (Fiorentino and Claps 1992; Fiorentino et al. 1992) show that the maximum entropy of the distribution of channel link elevations is a function of network diameter (D) such that

$$H = \ln D. \tag{5.9}$$

H is entropy and the development is based on the probability (p_i) that a network link will have the elevation E_i. A link is a stream segment connecting two nodes, where a node is either an external headwater source or a stream junction. The topological network diameter D is the maximum number of links between the basin outlet and a headwater source node. The maximum entropy in equation 5.9 will be reduced by the introduction of additional constraints, but the observed entropy is a function of, and is well predicted by, $\ln D$ (Fiorentino and Claps 1992; Fiorentino et al. 1992).

This suggests that the growth of any branching network increases entropy. Headward extension with no branching would not increase D; nor would stream capture or other mechanisms whereby tributaries are lost. However, headward extension with branching or bifurcations and additions of tributaries would increase D and thus H. While Fiorentino and others (Fiorentino and Claps 1992; Fiorentino et al. 1992) frame their discussion in terms of statistical entropy, we have already seen that statistical entropy measures the K-entropy of a nonlinear dynamical system such as a fluvial system. Therefore the growth of branching networks is chaotic, which would be expected to produce pseudo-random patterns, self-organization, and fractal geometry. At least some network and basin evolution models have been explicitly shown to be sensitively dependent on initial conditions (Willgoose et al. 1991b; Ijjasz-Vasquez et al. 1992; Rinaldo et al. 1993; Howard 1994; Moglen and Bras 1995).

Thus, we have a highly generalized proposed explanation for model equifinality in channel networks: the growth of a branching network is chaotic. This produces self-organizing, psuedo-random patterns with fractal geometry. Thus any model which results in the growth of a branching network is likely to produce networks which are visually, statistically, and topologically similar to real networks developed by headward growth and bifurcation. Downslope growth and coalescence of rills, and local downslope coalescence of channelized flows in the vicinity of channel heads and banks, does occur, and are indeed critical components of some network and basin evolution models. However, the dominant modes of growth once channels are established are indeed headward erosion and branching and lateral branching (Ruhe 1952; Schumm 1956; Knighton 1984).

This approach does help us to understand the model equifinality problem with respect to channel networks, but does not help in linking fluvial morphometry to processes, evolutionary history, or environmental controls.

Geological constraints

The fractal dimension of channels and networks can be interpreted (as can traditional morphometric relationships) in terms of geological constraints. High fractal dimensions can arise from network development in an area with few or no geological constraints on where channels can form or flow; low dimensions will be associated with higher degrees of lithological, structural, and tectonic constraints on channel locations (Phillips 1993g). In this section we use an NDS approach to further explore this argument.

Let the drainage basin be divided into n appropriately sized cells. Each cell has a value for a categorical response variable representing the stream network: for instance, unchannelled cells, cells with channels but no junctions, and cells with channel junctions. Each cell also has values for one or more categorical variables which represent controls on channel initiation and development. These control variables might represent classes of substrate erodibility (Stark 1991), stability thresholds (Kirkby 1971; Smith and Bretherton 1972; Tarboton et al. 1992), channel initiation functions (Willgoose et al. 1991a; Dietrich et al. 1992), other criteria used in network evolution models, or other constraints of interest or relevance to a particular problem or location.

If any value or outcome of the response or control variables, or combination thereof, could occur in any cell, then the entropies are at a maximum, corresponding to a random, white noise process. If the pattern is completely deterministic (and nonchaotic), then the channel response state of each cell is completely controlled (i.e. there is only one possible outcome given the state or value of the control variables), and the Kolmogorov entropy is zero.

Finite positive entropy is accounted for by colored noise and/or deterministic uncertainty. How much of this entropy is linked to environmental factors not accounted for (hydrologic response, vegetation, human agency, and so on) and how much to underlying constraints (presumably geological controls) which directly control or influence the control variables in the model? Referring to equations 2.25 and 4.10 and the arguments that precede them,

$$H_c(k) = -\ln\xi \qquad (5.10)$$

where $H_c(k)$ is the deterministic uncertainty (chaotic) component of the entropy, and ξ in this case represents the degree of geological constraint. The chaos or deterministic uncertainty is then a direct nonlinear function of the degree to which geological controls over factors such as substrate and topography determine where channels can exist. A high degree of control does not necessarily indicate greater overall complexity and uncertainty, but does show that a greater proportion of the observed complexity and uncertainty is deterministic.

This supports the notion that fractal and multifractal structures, topological measures, and other network morphometric indices are most reasonably interpreted as reflections of the nature and extent of geological constraints, rather than as indications of the space-filling, energy dissipation, or other apparent behaviors of fluvial systems. There are indeed systematic relationships between morphometry, landforming processes, and landscape evolution. However, because of the tendency of landforms and ESS outputs to develop very similar structures and outcomes independently of their processes and history, morphometry alone cannot allow one to deduce much about mechanisms or history (Culling 1957, 1987; Schumm 1991; Beven 1996; Phillips 1997a). To gain insight into processes and evolution, one must study processes and evolution.

Evolution of Topographic Relief

This problem encompasses a hoary yet essential question in the study of landscape evolution: does relief increase or decrease over time? There is insufficient space to recount the heritage of the relief question here, but the twentieth-century treatments began with, and were often directly or indirectly stimulated by, William Morris Davis's (1899) theory of cyclic landform evolution whereby relief is progessively lowered until uplift initiates a new cycle. The evolution of relief is important not just for the geomorphologist's efforts to explain landscape development, but also for reconstructions of past landscapes and for questions involving the extent to which landforms, stratigraphic units, and other landscape properties may be preserved. The analysis in this section represents a summary and updating of the more detailed analysis in Phillips (1995b).

As a general proposition, declining relief over time implies stability and nonchaotic, non-self-organizing evolution. As differences in elevation are reduced rather than magnified, the ever-declining range of elevations is progressively less sensitive to initial elevation differences. Increasing relief implies the opposite: as initial elevation differences, on average, increase over time, unstable, chaotic, self-organizing behavior is implied (plates 5.2 and 5.3).

After a brief survey of what major landscape evolution theories have to say about changes in relief amplitude, a theoretical framework is presented that explicitly links elevation change with the mathematical properties of nonlinear dynamical systems. Finally, the stable/nonchaotic and unstable/chaotic modes of topographic evolution are identified.

Theories of landscape evolution

Davis's "cycle of erosion" (Davis 1889, 1899, 1902, 1909) is often equated with declining relief due to the emphasis on "progessive lowering." However, the Davisian cycle also includes periods of increasing relief (figure 5.4). Rapid uplift initiates erosional cycles, marked originally by landscape dissection and valley deepening, i.e. increasing relief. Mass redistribution via erosion and deposition then leads to a progressive lowering of elevations and slope gradients, and ultimately to a low-relief peneplain. While progressive lowering and peneplanation are key elements of Davis's cycle, the early stages of uplift allow for increasing relief. So-called neo-

Plate 5.2
A fluvially dissected landscape near Billings, Montana, USA. Fluvial dissection is a divergent, relief-increasing mode of landscape evolution.

Davisian views which depict relief as a function of the antagonistic trends of uplift versus denudation (Scheidegger 1987) even more clearly recognize both increasing and decreasing relief as viable evolutionary trends.

The theory of landform evolution of Walther Penck (1924) accommodated the idea of low-relief endforms, but did not hold that they become initial forms for new episodes of dissection, as in Davis's cycle. Penck's scheme posits long phases of initial slow uplift. A new surface of low relief is developed, with an approximate steady-state balance between denudation and uplift. Once formed, slopes retreat parallel to themselves rather than downwasting, and relief is roughly constant. As uplift waxes to a maximum before waning to a minimum, dissection and relief amplitude increases can occur. Penck's system thus allows for increases, decreases, or relative constancy of relief during various stages of evolution.

L. C. King's (1953) theory of landscape evolution emphasized parallel slope retreat and retreating pediments. The consumption of

Plate 5.3
Soil redistribution by a combination of tillage, fluvial, and wind processes is resulting in net erosion of hilltop convexities and accretion of toeslopes and field edges in this agricultural landscape at Clayroot, North Carolina, USA. This is a convergent, declining-relief form of evolution.

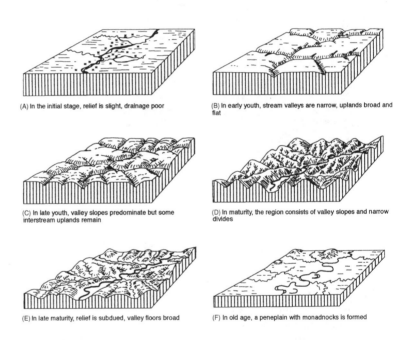

(A) In the initial stage, relief is slight, drainage poor

(B) In early youth, stream valleys are narrow, uplands broad and flat

(C) In late youth, valley slopes predominate but some interstream uplands remain

(D) In maturity, the region consists of valley slopes and narrow divides

(E) In late maturity, relief is subdued, valley floors broad

(F) In old age, a peneplain with monadnocks is formed

(G) Uplift of the region brings on a rejuvenation, or second cycle of dendation, shown here to have reached early maturity

Figure 5.4
The cycle of erosion proposed by W. M. Davis (1899). Relief increases in stages A through D, and decreases in D through F.

higher, older landscapes by scarp retreat is followed by subsequent erosion of the pediments thus formed by the scarp retreat of a younger and lower generation of pediments. The King theory is not based on a sequence of landforms, therefore, and implies dynamic but more or less constant relief.

C. R. Twidale (1991) developed a model whereby initial differences in weathering and erosion susceptibility become magnified over time and enhanced by denudational processes. Regolith development enhances the ever-increasing differences, leading to a steady increase of relief over time. He presents geological evidence to show that, in varied tectonic settings, relief has increased over periods of 60–100 million years. Crickmay's (1976) "hypothesis of unequal activity" also posits increasing landscape relief over time, in this case attributed to spatial variability in erosion due to differences in the erosive capacities of streams.

The theory of dynamic equilibrium (Hack 1960), actually based on a concept of steady-state equilibrium (Thorn and Welford 1994), suggests that relief might either increase or decrease as the landscape seeks an adjustment between the mass available for transport and the energy to transport it. The result, however, is a steady-state landscape of roughly constant relief, though it is recognized that considerable periods characterized by changing relief might be required to achieve that "dynamic" (steady-state) equilibrium.

Trends of increasing, decreasing, and relatively constant relief in the same landscape at different times are also present in numerical models of landscape evolution. Frank Ahnert's (1967, 1976, 1987, 1988) model of hillslope and landscape evolution is a mathematical simulation based on the mechanics of debris production, removal, and transport. Whatever the assumptions as to initial conditions or dominant processes, the general results with respect to the evolution of relief in the model are similar: relief first increases over time, steeply in the beginning and then at a slower rate; the rate of change in relief finally levels off and relief then begins to decline (figure 5.5). Armstrong's (1980) model of soil and slope evolution behaves in a broadly similar fashion: the difference between maximum and minimum heights above base levels increases at first, and eventually levels off. Brunsden's (1990) data from Taiwan suggests that the relaxation time for denudational adjustment to the increasing relief associated with uplift is about two million years. This supports the sequence of increasing, slowly changing, and then decreasing relief as landscapes evolve.

Chase (1992) simulated large-scale landscape evolution and topography in a numerical model, concluding that erosion "roughens" topography at all spatial scales, and thus presumably increases at least local relief. This is offset by slope-dependent diffusional processes at shorter, and deposition at longer, scales, both of which tend to smooth or decrease local relief. The change in relief may thus be a function both of the balance between roughening and smoothing processes and the spatial scale involved.

Ahnert (1984) has explicitly addressed the height and relief of mountain ranges as a function of the interaction of uplift and denudation rates. When uplift exceeds denudation, relief increases. Greater relief stimulates increases in erosion and denudation rates, but there is a lag time directly related to the areal extent of the uplift. The latter is attributed to denudation beginning at the edges

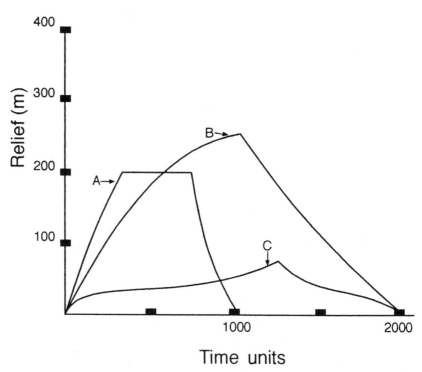

Figure 5.5
Trends in hillslope relief in Ahnert's (1987, 1988) simulation model. Despite differing assumptions about dominant mass transport mechanisms, curves A, B, and C all show a period of initially increasing relief and ultimately declining relief.

of the uplifted areas and moving toward divides by headward extension.

Adrian Scheidegger (1987, 1990) also attributes increasing or decreasing relief to the outcome of the interplay between endogenic (uplift) and exogenic (denudation) forces in his "principle of antagonism." Scheidegger (1983, 1987) also suggests an instability principle whereby relief variations or disturbances may persist or grow at certain scales. Both principles allow for either increasing or declining relief, depending on the relative strength of endogenic or exogenic forces (antagonism) or on the scale (instability).

Models and theories of landscape evolution, individually and in the aggregate, allow for increasing, decreasing, and relatively constant relief as the landscape evolves. Davis's cycle does emphasize downwasting and peneplanation, but clearly recognizes a stage of increasing relief. Twidale's and Crickmay's conceptual models call for increasing relief, but are plainly intended as "alternative" frameworks for landscapes where downwasting or steady-state relief is not observed. The Penck and King models, though best known for constant relief and parallel slope retreat, both allow for stages or situations involving increasing relief. Hack's steady-state ("dynamic") equilibrium implies roughly constant relief, but landforms in disequilibrium would exhibit changing relief while proceeding toward the stable end-state. Ahnert and Brunsden allow for a progression including increasing, steady-state, and decreasing relief, all of which could be interrupted at any time by tectonic or base-level changes. Scheidegger and Chase take this flexibility even farther: any type of relief change might occur, depending on the balance of forces or processes involved and/or the spatial or temporal scale.

The point of this brief overview is to show that various models and theories of landscape evolution allow for more, less, or approximately constant relief over time. The individual schemes are far richer and more complex than implied here, and you should consult the original sources to learn more about them.

Exponents and elevation

Does topographic relief increase or decrease as the landscape evolves? Consider the problem as a question of sensitivity to initial conditions: does convergence or divergence of elevations occur within the landscape? On average, do initial elevation differences decrease or increase over time?

Before returning to topography, consider a small sphere of n dimensions. As time marches on, the sphere evolves into an ellipsoid whose principal axes contract or expand at rates given by the spectrum of Lyapunov exponents λ_i. The λ may be arranged from the most rapidly expanding to the most rapidly contracting, so that

$$\lambda_1 \geq \lambda_2 \geq \ldots \geq \lambda_n$$

Recall that when the system's attractor is chaotic, the system trajectories in phase space diverge, on average, at an exponential rate given by the largest exponent (λ_1) (Eckmann and Ruelle 1985).

If you have all the equations describing the nonlinear dynamical earth surface system at hand, you can calculate the entire Lyapunov spectrum (Wolf et al. 1985). Usually, however, the geoscientist has field evidence and no complete set of equations. Fortunately, a convenient property of NDS is that the dynamics of the whole system can be deduced from a single observable (such as elevation).

In chapter 2 we noted that Rosenstein et al. (1993) showed that randomly chosen pairs of initial conditions in a chaotic system will diverge exponentially (in a stable system there would be exponential convergence) at a rate given by the largest Lyapunov exponent:

$$d(t) = Ce^{\lambda_1 t} \tag{5.11}$$

where $d(t)$ is the average divergence of randomly selected pairs at time t and C is a constant normalizing the initial separation. In our problem, $d(t)$ is the average divergence (or convergence) of elevation of pairs of randomly chosen sites at time t, and C normalizes the initial relief. If $\lambda_1 < 0$ there is increasing relief and topographic complexity (and vice versa!). The larger the value of λ_1, the more rapidly relief grows. We can see now that λ_1 actually depends on mean rates of elevation convergence or divergence by reiterating equation 2.13:

$$\lambda_1 = \ln d(t) - \ln C$$

If elevation differences grow over time, λ_1 becomes larger and must be positive. Decreasing relief would lead to smaller, negative λ_1s.

Let's get specific and consider any two points in the landscape, i and j, with elevations h_i and h_j, at times t and $t + \Delta t$. The elevation difference is given by $h_i - h_j$, and

$$
\begin{aligned}
d_{ij}(t + \Delta t) &= [h_i(t + \Delta t) - h_j(t + \Delta t)] \\
&= \{[h_i(t) + \Delta h_i] - [h_j(t) + \Delta h_j]\}
\end{aligned}
\tag{5.12}
$$

λ_1 is positive, and chaos exists, where

$$\Sigma d_{ij}(t + \Delta t) - \Sigma d_{ij}(t) > 0 \tag{5.13}$$

Interpretation

Landscapes are described below in terms of two locations corresponding to i and j, as a convenient way of describing the mean response of randomly selected pairs of locations. If two locations are initially equal in elevation, evolution will be stable and nonchaotic only if $\Delta h_i = \Delta h_j$. Locations with initially unequal elevations will experience declining relief and stable, nonchaotic evolution under four other conditions. This leads to five stable modes or situations: modes 1–5 in table 5.2. Unstable, chaotic evolution with increasing relief will occur with initial points of equal elevation if any inequity in Δh occurs between them. Otherwise, increasing elevation differences will occur in four other situations. Thus the five modes of chaos shown table 5.2.

Topographic evolution

Does relief tend to increase or decrease over time? Clearly, the answer is yes. A single landscape may undergo phases of both advancing and declining (and relatively static) relief, and examples of each abound in the contemporary landscape and the geological record. Rather than emphasizing one trend in relief, or a particular progression of trends, this NDS-based view of topographic evolution views the problem as one of whether elevation differences, on average, converge or diverge over time. This approach reveals that:

1 There are ten fundamental modes of topographic evolution, based on the average rates of uplift/accretion or erosion/subsidence at initially higher and lower sites. Five

Table 5.2 Ten general modes of relief evolution based on the average convergence or divergence of elevation between randomly selected locations (modes 1 through 5 are stable and nonchaotic; 6 through 10 are unstable and chaotic)

Mode	Initially higher site[a]	Initially lower site[a]	Generic example
1	Initial elevations equal; rates of elevation change equal		Planar surface with spatially uniform erosion, accretion, or uplift
2	Eroding; ≥lower site	Eroding; ≤higher site	Slope wasting, where mass removal on the upper slope is greater than on the lower slope
3	Uplifting/ accreting; ≤lower site	Uplifting/ accreting; ≥higher site	Uniform regional uplift, accompanied by deposition at lower sites to create more rapid elevation increases there
4	Eroding	Accreting/ uplifting	Sediment eroded from upper slope and deposited at slope base
5a	No change	Accreting/ uplifting	Stable interfluve above an aggrading valley
5b	Eroding	No change	Pediment beneath retreating slope face
6	Initial elevations equal; rates of elevation change unequal		Plane with any variation in erosion, deposition, or uplift rates
7	Eroding; <lower site	Eroding; >higher site	Stream incision which exceeds erosion rates of valley walls
8	Accreting/ uplifting; >lower site	Accreting/ uplifting; <higher site	Block faulting
9	Uplifting/ accreting	Eroding	Fluvial dissection during epeirogeny
10a	No change	Eroding	Stream incision below stable upland interfluve
10b	Uplifting/ accreting	No change	Shore platform below accreting beach

[a] The initially higher and lower sites refer to any two sites in the landscape.

modes are stable and nonchaotic; five are unstable and chaotic (table 5.2).

2 Stable modes feature constant or generally declining relief over time. Unstable, chaotic modes are characterized by increasing relief.

3 Neither chaotic nor stable modes can persist indefinitely, over geological time scales. Erosion, downcutting, and subsidence must eventually be limited by base levels. Uplift or accretion must eventually be limited by achievement of isostatic equilibrium, exhaustion of mass, or aggradation to limiting elevations. This means relief divergence or convergence must eventually reach some limit and halt or reverse itself. Thus the long-term persistence of any mode must ultimately result in a shift to a different mode.

4 Chaotic modes of topographic evolution involve sensitivity to initial conditions and inherent unpredictability. However, the increasing topographic complexity and the unpre-

dictable response in detail occur within well-defined boundaries associated with relative rates of uplift or accretion and erosion or subsidence.

5 Planar surfaces are unstable, and sensitive to any anisotropy in rates of uplift, subsidence, erosion, or deposition.

As noted previously, existing theories of topographic evolution can be mapped on to one or more of the ten basic modes, and each mode can be interpreted in terms of one or more existing theories (Phillips 1995b). However, this approach does not represent an alternative or meta-theory, precisely because it can accommodate any observed or inferred trend in the evolution of topography and relief. It is unfalsifiable. What the analysis does tell us is this:

- Some earth surface systems have both stable, nonchaotic, non-self-organizing, and unstable, chaotic, self-organizing modes. Neither mode is necessarily more common, important, or "normal" than the other.
- There may be inherent limits on particular modes of ESS development, whereby neither stable nor unstable behavior can continue indefinitely, and both are inevitable over long time spans.

Desertification

Thomas and Middleton (1994) make a strong case that the dimensions and severity of dryland degradation problems (desertification) have been strongly overstated from the 1970s into the 1990s. However, even the somewhat smaller and more restricted scope of the desertification problem outlined by Thomas and Middleton (1994) and reflected in estimates from the 1990s (Middleton and Thomas 1992; Williams and Balling 1996) still amounts to dramatic environmental change with serious economic and land-use implications and in some cases, a terrible human cost.

Desertification definitions, dimensions, processes, and problems are thoroughly reviewed elsewhere (Barrow 1991; Mainguet 1991; Middleton and Thomas 1992; Thomas and Middleton 1994; Williams and Balling 1996). This form of land degradation is an important and interesting focus for analysis of ESS for several reasons. First, the environmental changes involve interactions between climate, biota, soils, hydrology, and geomorphic processes, and may be initiated by direct or indirect, natural or anthropic modifications of any of the above. Second, desertification studies have identified critical issues directly related to system stability, including inherent instability and complex nonlinear dynamics in dryland climates (Wright 1980; Lockwood 1986; Demaree and Nicolis 1990), in soil moisture–climate interactions (Rodriguez-Iturbe et al. 1991), in semi-arid vegetation and ecosystems

(Westoby et al. 1989; Freidel 1991; Laycock 1991; Tausch et al. 1993; Walker 1993; Hobbs 1994), and in the broad interactions between climate, vegetation, runoff, erosion, and soils (Thornes 1985, 1988; Phillips 1993b; Phillips and Renwick 1996). More generally, the role of feedbacks in dryland environmental systems and the persistence of environmental changes in drylands have emerged as critical research and policy issues. In fact, much of the research on dryland ESS in the past two decades stems from seminal studies linking surface changes due to vegetation and soil disturbance to climate and hydrological changes (see Williams and Balling's 1996 synthesis).

This section, building on Phillips (1993b), seeks to address the basic question of whether there are fundamental, ubiquitous biophysical feedbacks or instabilities in dryland environments which make them inherently vulnerable to degradation or, more generally, to disproportionately large or long-lasting environmental changes in response to perturbations. Beyond the general prevalence of instability and deterministic uncertainty in ESS (see chapters 3 and 4), there are at least three other good reasons to suspect that this may be the case. First, even the downward-revised estimates of dryland degradation of the past few years show significant desertification on every continent except Antarctica, in a variety of environmental contexts, and under many different land-use and management schemes (Middleton and Thomas 1992; Williams and Balling 1996). This suggests that common, fundamental properties of dryland ESS rather than (or in addition to) regional and local factors may be involved. Second, in at least some cases the degradation or environmental change is persistent or irreversible (over human time scales). This raises the question of whether a degraded state may be an attractor or one of several multiple equilibria. Finally, several analyses of specific desertification patterns and processes, in both models and field studies, indicate instability, deterministic chaos, and related behaviors such as divergence from similar initial conditions and increasing divergence over time (Thornes 1985, 1988, 1990; Schlesinger et al. 1990, 1996; Phillips 1993b; Abrahams et al. 1995; Savenjie 1995; Parsons et al. 1996; Puigdefábregas and Sánchez 1996).

Biophysical feedbacks model

Schlesinger et al. (1990) presented a model illustrating the biophysical feedbacks involved in semi-arid land degradation, which represents a broad consensus of at least the qualitative nature of the interrelationships (Williams and Balling 1996). I modified the model (see figure 1.2) to depict the critical relationships among system components independently of the evolutionary trend of the system at any given moment, and added four links, two of which reflect effects of aeolian erosion on soil heterogeneity and of climate change on runoff. The other two are self-effect loops reflecting climate persistence independent of interactions with the land surface, and vegetation/organic matter (as expressed in the organic nitrogen component) density dependence. A more complete explanation is given in an earlier paper (Phillips 1993b).

The system has already been shown to be unstable by the Routh–Hurwitz criteria (Phillips 1993b; Phillips and Renwick 1996). Further, when the various bivariate cause-and-effect relationships between vegetation cover, albedo, temperature, precipitation, soil moisture, aeolian erosion, and water erosion identified in the desertification literature are examined, the majority promote instability in response to perturbations (Phillips 1993b). Rather than re-hash the analysis, let us further examine the implications and subsequent field evidence relevant to them.

Implications

Analysis of the interrelationships and feedbacks involved in desertification indicates an inherently unstable, potentially chaotic, self-organizing system. This suggests that small perturbations from any source, be they natural climatic cycles, volcanic eruptions, overgrazing, groundwater withdrawal, or whatever, will tend to persist and grow over time. Further, the changes in the dryland environmental system will be disproportionately large or long-lived compared to the magnitude and/or duration of the disturbance. This finding supports the notion that many dryland systems are inherently vulnerable to both human and non-human disruptions. The robust approach inherent in the qualitative stability analysis is important here. It may be problematic to apply the results of a location-specific field or modeling study to other locations, even quite similar ones. But, though the specific, quantitative response of wind erosion or runoff to vegetation removal, for example, may be quite different in Mexico, Kuwait, and Botswana, or at different times in the same place, the general, qualitative fact that removing vegetation enhances aeolian erosion and runoff holds true everywhere and all the time. The relevance is not diminished by the fact that this relationship is on an other-things-being-equal basis, as the other things are accounted for.

Management implications are discussed elsewhere, with respect to the biophysical feedback model in particular (Schlesinger et al. 1990; Phillips 1993b; Phillips and Renwick 1996; Williams and Balling 1996) and the phenomena of unstable rangelands with multiple equilibria in general (Westoby et al. 1989; Freidel 1991; Laycock 1991; Tausch et al. 1993; Walker 1993; Hobbs 1994). With respect to ESS behavior, results imply that:

- Perturbations to dryland ESS, whether natural or human, and whether positive or negative, are likely to persist and grow over the short term. Environmental changes are likely to be disproportionately large relative to the external forcing. It then follows that:

Plate 5.4
Erosion features often persist in degraded drylands. These exposed subsoils are associated with eroded trails in Phoenix, Arizona, USA.

- Dryland ESS are self-organizing, and should exhibit average trends toward greater spatial heterogeneity over time in terms of the patchiness and spatial variability of factors such as soil properties, vegetation cover, moisture and hydrologic response, and erosion–transport–deposition patterns.
- The general instability leads to situation-specific, testable predictions. To give three examples, instability leads to the hypotheses that: (a) there will be increasing divergence of the landscape into soil-resource-rich and soil-resource-poor patches, once initiated by grazing pressure or shrub invasion; (b) changes in local moisture recycling initiated by soil or vegetation changes will persist in the form of local rainfall reductions (or increases); and (c) the initiation of local soil erosion will result in a self-reinforcing increase in erosion rates and the development of semi-permanent erosion features such as gullies, erosion pavements, or exposed calcic horizons (plate 5.4).

Empirical evidence

What does the empirical literature tell us about the implications and predictions above?

Patch divergence   Overgrazing and shrub invasion do result in an increasingly differentiated spatial mosaic of soil nutrients and moisture. Studies of energy and water budgets of patterned woodlands in the Sahel, for example, show that bare soil areas act as rainfall collectors for vegetation patches, thus enhancing and reinforcing a self-organizing pattern of drier, bare soil and moister, vegetated soil (Culf et al. 1993). In the deserts of the south-western US, "islands of fertility" develop under shrubs as semi-arid grasslands are invaded by shrubs. Over time, there is a shrub-intershrub divergence of nutrient-rich and nutrient-poor soil (Schlesinger et al. 1996). Abrahams and others (1995) found that the grass-to-shrub vegetation change at Walnut Gulch, Arizona, increases inter-rill erosion and runoff and the spatial heterogeneity of the plant cover. This in turn leads to further variability in desert pavement formation, soil fertility, erosion, and surface hydrologic responses. Puigdefábregas and Sánchez (1996) also documented patch divergence in semi-arid Spain, and linked vegetation and soil patchiness to water and soil circulation.

In the Negev desert, Yair (1990) found that the development of a spatial mosaic of bedrock outcrops and soil-covered areas leads to ever-increasing divergence in regolith development. Runoff from the rock outcrops concentrates water, infiltration, leaching, and associated weathering in soil-covered areas, and also leads to salinity concentrations.

Moisture recycling   While the specifics are subject to considerable debate and uncertainty, it is increasingly clear that the persistence of wet/dry conditions is linked, at least in part, to land surface–atmosphere feedbacks (Lare and Nicholson 1994; Williams and Balling 1996). The persistence of minor or short-lived perturbations in moisture recycling is illustrated by Savenjie's (1995) work on land use, vegetation, and precipitation interactions in the Sahel. Vegetation destruction or reduction must inevitably lead to decreases in transpiration and precipitation and increases in the proportion of precipation running off, he found, which in turn reinforces the vegetation decline. Xue and Shukla (1993), also working in the Sahel, compared model results of the influence of land surface properties on climate during desertification with field data. They found that as degradation intensifies, rainfall declines and the rainy season is delayed due to feedbacks from vegetation to soil depth, soil moisture, and albedo, and from these variables to precipitation.

Lapenis and Shabalova (1994) developed a model accounting for feedback between precipitation and development of a root zone, arguing that such feedback is responsible for a long-term response of local hydrologic cycles to global temperature change. Thus small or gradual changes in climate or vegetation can lead to fundamental transitions between tropical forest, savannah, semi-arid, and desert vegetation formations, taking about 1500 years to reach a new steady-state equilibrium. The implications of the Lapenis and Shabalova (1994) model are supported by paleoclimatic data from Africa.

Erosion features Self-organized growth of erosion–transport–deposition sequences at a variety of scales in semi-arid Australia has been documented by Pickup (1988; see also Pickup and Chewings 1986). The persistence and growth of erosional features in response to small perturbations is evident in Barth's (1982) studies of fossil dunes in Mali. Vegetation removal and trampling initiates both water and wind erosion. Once begun, Barth found that both processes are self-reinforcing. Experiments at Walnut Gulch, Arizona (Parsons et al. 1996) show that minor vegetation change has had substantial geomorphic impacts. Compared to grassland, inter-rill portions of hillslopes invaded by shrubs have higher runoff rates, more frequent steady-state runoff, greater overland flow velocities, and higher erosion rates.

The question of scale

Despite ample evidence of the inherent instability of dryland earth surface systems, it is not yet clear that an instability-based paradigm for understanding and managing these systems is appropriate under all circumstances. Thomas and Middleton (1994), for example, present abundant evidence that some cases of vegetation change once interpreted as desertification were more likely simply natural responses to short- and medium-term climate change, and that many supposedly degraded areas recovered when the rains returned. Perkins and Thomas (1993) show that in Botswana vegetation is vulnerable to disturbance but recovers quickly afterward (though other factors recover less quickly). Baker and Walford (1995) suggest, based on their work in the western US, that state-and-transition models based on paradigms of instability and multiple stable states are less appropriate than succession-based models which account for disturbances and ecological mosaics. They also argue that multiple stable states are uncommon in natural ecosystems, and that the plethora of observed stable ecosystem states in human-influenced agro-ecosystems are symptoms of failed management and poor rangeland health, not the inherent response to disturbances.

Ecosystem recovery in response to wetter climates does not necessarily invalidate the notion of inherent instability, as it can be argued that the system is unstable in response to any change in boundary conditions. Therefore the desertified state, unless degradation has reached the point that vegetation and soil recovery is impossible, is just as unstable as the non-degraded state, and thus prone to change states when the rains return. But this does raise troubling issues regarding scale. What constitutes an unstable, perhaps reversible state-change, as opposed to a transient disturbance to a stable system? The answer almost surely varies with spatial and temporal scale.

The fundamental principle of physical geography that the controls, and even the nature, of process–response relationships vary with spatial and temporal scale is certainly true of desertification and dryland ESS. For example, the factors controlling runoff response in semi-arid Spain vary from the patch level, where organic matter, soil hydrophobicity, and individual plants control the hydrologic response, to the small slope or runoff plot scale, where spatial patterns of vegetation, bare ground, preferential flow paths, and permeability control the response (Nicolau et al. 1996). Seyfried and Wilcox (1995) obtained broadly similar results over a wider range of scales in the western US.

With respect to temporal scales, most desertification studies have been (appropriately enough, given the applications to environmental management) over scales of decades or shorter. There is good reason to suspect that studies over longer time scales will yield at least somewhat different results and implications. Over Quaternary time scales, Yair (1992) presents geomorphic evidence of complex climate and surface process interactions along the fringes of the Negev desert. The dominant aeolian sediment input varies with climate, and is mainly loess (silt) in wet periods and sand in dry cycles. Loess deposition over rock reduces runoff due to increase soil moisture storage, with shallower infiltration and greater evapotranspiration losses. Thus there is an increase in soil aridity, even in a moister climate! Loess deposition over sand, however, stabilizes dunes and reduces runoff and erosion.

The question of spatiotemporal scale, and its relationship to deterministic uncertainty and nonlinear dynamics, is one we must confront in subsequent chapters.

6

Soil: the Prototype Earth Surface System

As far as human perception and comprehension are concerned, there is an infinite number and variety of earth surface systems. Further, there are important interrelations between the atmo-, litho-, hydro-, and biospheres in a variety of earth surface forms, processes, and phenomena which may produce outcomes reflecting those multiple influences and complex interactions. With this in mind, I propose here that the soil and regolith represent a canonical earth surface system. By this I mean that soils and regoliths, as a generalization, provide the best example of an ESS reflecting the influences of all the environmental spheres. Further, soils can provide an excellent record of environmental changes and influences over a broad range of time scales. As soils also represent ESS which exhibit complex nonlinear interactions and extreme local spatial variability, it can be argued that they are the prototype ESS.

The State-factor Model

The interdependence of soils, climate, geology, topography, and organisms has long been recognized, but scientific ideas on the interrelationships began to flower in the late nineteenth century. Russian geologist, geographer, and pedologist V. V. Dokuchaev was the most important protagonist. He asserted in 1879 that soils are not simply a geological formation or surficial geological debris; rather, they are independent bodies including both mineral and organic substances and produced by the combined activity of animals, plants, climate, relief, and geology (Joffe 1949: 17). This is the so-called factorial or state-factor model: soils or soil properties are viewed as a function of soil-forming factors. Dokuchaev's was the first definitive statement of the factors of soil formation, and provided both a conceptual and an analytical framework for understanding soils as products of the environment (and, later, for inter-preting environmental conditions from pedologic evidence). As these ideas were developed and disseminated by Dokuchaev and his students and colleagues in the late nineteenth and early twentieth centuries, roughly parallel conceptual frameworks were being devised in ecology and geomorphology (Johnson and Watson-Stegner 1987; Huggett 1995: 28–9; Osterkamp and Hupp 1996).

Several major elaborations of the state-factor model occurred in the early twentieth century. This culminated in the most famous and influential statement, Hans Jenny's (1941) book *The Factors of Soil Formation.* In it, Jenny worked out the functional nature of environmental influences on soil, and produced the "clorpt" equation:

$$S = f(cl, o, r, p, t) \ldots \tag{6.1}$$

where S is the soil or a particular soil property, and the other variables represent, respectively, climate, organisms (biosphere), relief (topography and drainage), parent material (geology), and time. The trailing dots represent state factors which may be important locally but not universally, and which cannot be interpreted in terms of the five standard state factors. Aeolian deposition is the most commonly provided example of a "dot factor;" extraterrestrial impacts or anthropically driven acid deposition might be others.

The purpose and scope of the state-factor model is sometimes misunderstood. The factors are not intended to describe pedogenetic processes or soil components. Rather, they provide the context and boundary conditions within which soil genesis and evolution occur. As Jenny (1961: 385) put it, "the factors are not formers, or creators, or forces: they are the variables (state factors) that define the state of the soil system" (figure 6.1). The approach is ecological and systems-oriented in that the soil cannot be considered in isolation from its physical, chemical, and biological environment (Jenny 1941, 1961, 1980).

The state-factor model complements process-oriented approaches to pedology. Paton (1978), who favors the latter, notes that state factors operate at broader spatial and temporal scales than processes, and set limits to the nature and rates of pedogenetic processes. Birkeland (1984) argues convincingly that the state-factor approach facilitates the isolation and study of critical pedologic and geomorphic processes in soil geomorphology. Though it is not unchallenged, and adherents do not claim it is superior for all pedologic problems, the factorial approach remains the dominant paradigm for pedology, soil survey and mapping, soil geomorphology, and paleopedology (Birkeland 1984; Retallack 1990; Hudson 1992; Johnson and Hole 1994; Wilding 1994).

There have been several efforts to generalize or broaden the pedologic state-factor model to other environmental systems. These are discussed in the next section. There have also been reinterpreta-

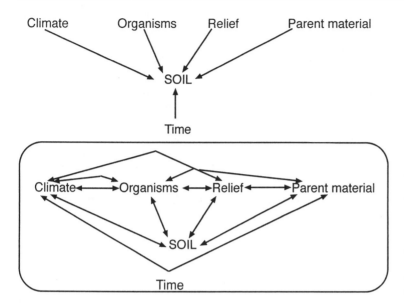

Figure 6.1
The original state-factor model, shown at the top of the figure, was concerned with soil as a product of the interactions between climate, biota, topography, parent material, and time. More recent versions, shown in the lower part of the figure, have incorporated time in dynamical systems models and explicitly consider the interactions between soil and the interdependent state factors.

tions and simplifications, such as those of Runge (1973) and Chesworth (1973). Both may be considered special cases of the factorial model, applicable respectively to unconsolidated surface deposits of equal weathering degree where leaching is the dominant process or to highly weathered rocks where parent material influences have largely been obliterated (Yaalon 1975).

Generalizations

The state-factor model was extended to cover entire ecosystems by Major (1951) and Perring (1958), such that not only soil, but biotic communities as well are interpreted as products of the state factors. The factorial model has also been adapted to geoarcheology (Holliday 1994) and to geomorphology and Quaternary geology (Johnson and Hole 1994). The most extensive factorial model is that of Pope et al. (1995) who sought to explain geographical variations in weathering as a function of 16 state factors ranging from lithology and structure to microtopography and biotic and abiotic reaction chemistry. Jenny (1961) himself reformulated the model in terms of fluxes of energy and matter in ecosystems, a view previously proposed by Nikiforoff (1959). These fluxes occur along gradients or potential differences such as hydraulic head, gravity, or gradients of chemical concentrations or temperature. If the system has thickness Δx the flux rate per unit cross-section is a function of the internal and external potentials (P_i, P_x) and a permeation parameter (m) which relates to the ease with which the material can pass through the system.

$$\text{flux} = -[(P_i - P_x)/\Delta x]m \qquad (6.2)$$

Equation 6.2 is a generalized function; Darcy's law for fluid flows through porous media, and laws describing thermal conduction and chemical diffusion are examples of specific forms.

Jenny (1961) introduced a notion of historical dependency which presaged a good deal of NDS literature. P_i and m in any soil or ecosystem, Jenny argued, are governed by the condition or state of the system at that moment. For example, soil moisture storage and hydraulic conductivity are linked to the degree of soil development at a given time – or, at a shorter time scale, both vary systematically through the course of a hydrologic event. For another example, the amount and quality of organic litter and the maximum rate of mineralization are functions of the stage of development of the ecosystem with respect to vegetation, litter, soil organic matter, and soil fauna. Then

$$\text{flux} = f(P_x, \Phi) \tag{6.3}$$

where Φ is the state of the system. If Φ is time-dependent, then

$$\text{flux} = f(P_x, \Delta t) \tag{6.4}$$

where Δt is the age of the system or elapsed time since the most recent disturbance. Jenny (1961) then proceeds to a generalized state-factor model of the form

$$L = f(L_0, P_x, t) \tag{6.5}$$

where L represents the ecosystem and/or soil properties, L_0 the initial conditions, and t the age or state of the system. Thus, Jenny's (1961) generalization of the state-factor model implies a view consistent with NDS theory: ecosystems and soils are time-dependent, sensitive to initial conditions, and evolve within ranges set by external environmental conditions.

The form of the "clorpt" equation is reminiscent of a multiple regression equation, leading to objections that the state factors are not independent. While Jenny (1941) did intend for the state-factor equation to provide the foundation for prediction of soil properties based on measurements or indicators of the state factors, the predictions need not be quantitative or based on assumptions or interpretations of the soil-forming factors as independent variables. The interdependency of the state factors was recognized early on (Stephens 1947; Chesworth 1973; Yaalon 1975), and was in fact acknowledged implicitly (Jenny 1941) and later explicitly by Jenny (1961, 1980; Amundson and Jenny 1991).

Equation 6.5 recognizes time as the only truly independent state factor, as both initial conditions (L_0) and external controls (P_x) may encompass a number of environmental controls. Thus (ignoring "dot factors" for the sake of clarity), we can use the subscript t to denote a particular time or temporal scale, and rewrite equation 6.1 as

$$S_t = f_t(cl, o, r, p) \tag{6.6}$$

Then an interaction matrix describing the interactions between soil, climate, organisms, topography, and parent material can be constructed with respect to a particular soil phenomenon, and the stability analyzed using the methods described in chapter 2 and deployed elsewhere in this book. Phillips (1989d) did this, and devised a broadly applicable interaction matrix which suggested that the full five-factor (s, cl, o, r, p) system is inherently underdetermined. This arises because at either short or long time scales climate and parent material components are independent of topographic, organism, and soil factors (this applies only to the most general form of the model, and often does not apply when the factors are defined with respect to specific soil characteristics and landscapes). In this context parent material was interpreted as the initial conditions (L_0), climate as the primary determinant of external potentials for mass and energy fluxes (P_x), and soil, vegetation, and topography as the internally interacting variables which change systematically as a function of time. A reduced system, where soil, organisms, and relief interact under conditions of constant climate and geology, was also found to be unstable (Phillips 1989d).

Richard Huggett (1991) took the factorial model a step further, in the context of relating the world climate system, biosphere, and other geosystems. In the most general case, the rate of change in any state variable is a function of all the other state variables, and a canonical nonlinear dynamical system consisting of a set of simultaneous differential equations of the style represented by equation 2.1 is obtained. Using the time rate of change rather than time as a state factor, Huggett (1991) adapted Jenny's model to derive an equation system, a development presaged by Stephens (1947). Later, Huggett modified the labeling to more explicitly indicate the interactions of the environmental spheres, thus transforming and generalizing the "clorpt" equation to the "brash" equations, whereby b is the biosphere, r the toposphere (relief and topography), a the atmosphere, s the pedosphere (soils), and h the hydrosphere (Huggett 1995):

$$db/dt = f(b, r, a, s, h) + z \tag{6.7a}$$

$$dr/dt = f(b, r, a, s, h) + z \tag{6.7b}$$

$$da/dt = f(b, r, a, s, h) + z \tag{6.7c}$$

$$ds/dt = f(b, r, a, s, h) + z \tag{6.7d}$$

$$dh/dt = f(b, r, a, s, h) + z \tag{6.7e}$$

Here z represents external driving forces such as heat from the earth's interior, volcanic ejecta, and extraterrestrial inputs. Further

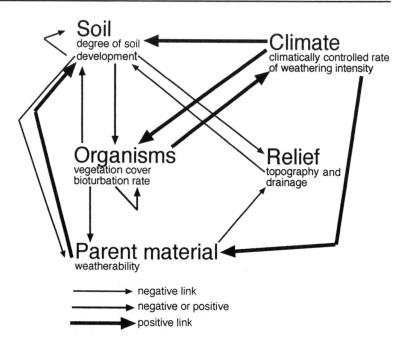

Figure 6.2
A diagram version of Phillips (1993d) reformulation of the state-factor model as a nonlinear dynamical system.

details, and discussion of how the environmental spheres are defined, are given by Huggett (1995). By placing the state-factor model for soils in an even broader context than Jenny (1961) had done, Huggett (1991, 1995) established a framework for examining soils (among other things) as indicators of earth surface system interactions in an NDS context.

Independently, and more or less simultaneously, I also reinterpreted the state-factor model as a nonlinear dynamical system, but more strictly related to soils, and with each component defined with respect to a particular soil property, the degree of soil profile development (figure 6.2; Phillips 1993d). The analysis, while generalized, was based on conditions typical of soil landscapes in my field area at the time, the North Carolina coastal plain, and showed that the soil system is unstable and potentially chaotic. Increasingly specific models in the same study area were linked to field-testable hypotheses which provided field evidence of chaotic soil evolution (Phillips 1993a; Phillips et al. 1996) and identified specific soil geomorphic processes which control the stability of pedologic evolution (Phillips 1996a).

Independently of explicit factorial models, Vitousek (1994) shows how the factorial approach, applied to ecosystem structure and function, explicitly treats state-factor interactions with each other and with processes internal to the ecosystem. He also argues that the development and validation of process-based ecosystem models is dependent on state-factor approaches, a view echoed by McSweeney et al. (1994) with respect to soil models.

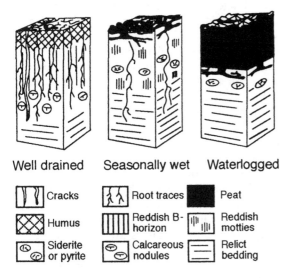

Well drained   Seasonally wet   Waterlogged

	Cracks			Root traces			Peat
	Humus			Reddish B-horizon			Reddish mottles
	Siderite or pyrite			Calcareous nodules			Relict bedding

Figure 6.3
Characteristic patterns of root traces, cracking, peds, nodules, color, and relict bedding in paleosols. The persistence of these features allows the interpretation of past water regimes (from Retallack 1990).

Summary

The state-factor model of soils views the latter as products of the environment, expressing the interactions of climate, geology, biology, hydrology, and other factors. Recent adaptations and applications of the approach have embraced the interactions between the state factors and soils themselves, rather than seeking to control or minimize them. The success of the state-factor approach in explaining soils, landscapes, and ecosystems argues strongly for a view of soils as a canonical ESS, and provides a conceptual framework for analysis in that context.

Paleopedology

A view of soils as a prototype ESS, reflecting the effects of climate, geology, hydrology, and the biosphere is, in essence, the underpinning of paleopedology. While the study of paleosols is sometimes driven by the desire to understand and explain soils of the past, more often the overriding purpose is to use fossil soils to make paleoenvironmental interpretations (figure 6.3). This essentially inverts the logic of the state-factor model, but paleoenvironmental interpretations of paleosols is an "enterprise dependent on the enormous literature on factors in soil formation" (Retallack 1994a: 31).

The view of soils as reflections of broader environmental conditions is underscored by the fact that most cautionary words about, and objections to, the use of paleosol features as paleoenvironmental indicators are based on the influences of multiple state factors. Kiernan (1990), using examples from glacial deposits in Tasmania, shows that weathering features can be used as reliable indicators of age only in cases where the influence of topography, organisms, and climate, which can obscure age-dependent relationships, can be minimized. Catt (1991) provides analogous advice with respect to soils as indicators of Quaternary climate change because the effects of climate and the timing of climate change are difficult to disentangle from parent material, biotic, and topographic effects. Verosub et al. (1993) show that paleomagnetic stratigraphy of loess sequences in China, previously interpreted as reflecting climate change, may reflect pedological processes instead of or in addition to climatic forcings. The complications that the interactions and multiple causality of environmental factors cause for interpretation of paleosols are discussed in detail by Retallack (1990) and summarized by Retallack (1994a), as is the general solution: the identification of specific relationships between soil properties and environmental controls and the conditions under which they hold. Two examples are frost features as indicators of periglacial climates in mid-latitude paleosols (Catt 1991), and the relationship between mean annual precipitation and depth to calcic horizons (Retallack 1994a).

As with the application of the state-factor approach to surface soils, the interactions of the atmo-, litho-, hydro-, and biospheres as expressed in the pedosphere can be confronted directly rather than circumvented. This is evident in Retallack's (1994b) view of "ecosystem paleopedology," which includes the concept of the pedotype. Pedotypes are paleosols which represent a trace fossil of an entire paleoenvironment.

Complex Interactions

Reviews of the influences of particular state factors on soil processes and properties are given by Birkeland (1984), Retallack (1990), Gerrard (1992), and Ollier and Pain (1996) for both modern and fossil soils. Here, it will better serve my argument to cite a few examples illustrating not only the polygenetic nature of soils, but the complex interactions between soils and other environmental factors.

Complex interactions between soils, plants, hydrology, and microclimate, for example, characterize ecosystems in general and montane ecosystems in particular. Kupfer and Cairns (1996) show how these interactions lead to disequilibrium, lags in adjustment, and complex response of montane ecotones to climate change, such that shifts of such ecotones are not a reliable indicator of climate at useful time scales. This is evident in the patchiness of plant communities and edaphic conditions in upper timberline environments in Colorado, which is attributable to a complex relationship between topography, vegetation, climate, and soil (Holtmeier and Broll 1992). Wind flow near the soil surface is strongly influenced by microtopography and tree islands. This in turn affects the distribution of snow cover, and thus soil moisture and soil processes.

Pennington (1986) suggests that otherwise inexplicable observed responses of vegetation to climate change in the UK could be explained by differentiation of soils. The more slowly responding soils influence trees, with which there are mutual interactions. In areas of variable parent material, vegetation patterns are likely to be most influenced by effects of soil development and the development of a soil mosaic rather than climate.

Recognition of the complex interrelationships of environmental factors has resulted in an increasing effort by biogeographers and ecologists to move beyond analyses of species distributions along environmental gradients. The simple up- and down-fan gradients described for desert alluvial fans, for example, may not hold up to closer scrutiny. Parker (1995) documents a complex network of relationships between geomorphic history, depositional surfaces, soils, and vegetation on Arizona fans. These interrelationships rather than simple age or elevation gradients determine the ecological patterns. McAuliffe (1994) also studied the integration of landscape evolution, soil formation, and ecological patterns and processes on Sonoran desert bajadas. His work stresses the importance of understanding ecological patterns in landscape context, and the role of soil and landforms in controlling ecosystems. This theme is extended to temperate riparian corridors, slopes affected by mass movements, and non-mountainous glaciated areas by Parker and Bendix (1996). Vegetation–landform feedbacks, typically mediated by soil, are shown to be critical, and there is considerable evidence of self-reinforcing mechanisms which enhance evolving soil and vegetation contrasts.

The distribution of plant communties in cerrados (savannahs) of central Brazil would, at first glance, appear to be controlled by a moisture gradient. However, vegetation and soils co-vary along moisture gradients, complicating the causal pathways. Further, the moisture fluxes are largely controlled by topography, though influenced significantly by soil properties and vegetation. Ultimately, the geo-ecosystem must be understood in terms of the interactions of topography, hydrology, soils, and vegetation (Furley 1996).

The complex interactions are also well illustrated by model-based studies. Podzol chronosequences from Scottish highlands show changes in saturated hydraulic conductivity over 13,000 years. Brooks and Richards (1993) use chronosequence data and a process model to show that as the soil has differentiated into three horizons (largely as a result of hydrologically driven processes), the resulting soil moisture storage and flux properties have changed. This is significant for modifying runoff processes and hillslope stability. A complex, irregular, nonsystematic pattern of soil depth was derived from a straightforward deterministic process-based model for the evolution of regolith-mantled slopes (Kirkby 1985). The interactions of lithology, slope hydrology, soil depth and formation rates, slope processes, and slope form produce the complex outcome, even along modeled hillslopes of uniform lithology and total denudation rate.

Coevolution

The coevolution of soils and landforms, linked as it is by weathering and mass flux processes, is reasonably obvious. Less obvious but equally important is the coevolution of soils and other environmental factors. The coevolution of soils and terrestrial ecosystems, for example, has been called one of the most significant outcomes of biological evolution (Richter and Markewitz 1995). It was this coevolution that inspired Major (1951) and Perring (1958) to extend the factorial model from soils to ecosystems. This coevolution can confuse the interpretation of cause and effect. In southern California chaparral, for instance, microscale soil heterogeneity is often cited as the cause of vegetation heterogeneity. But Beatty (1987) pointed out that most soil properties which vary systematically among vegetation types are those likely to be influenced by the shrubs themselvses.

An inextricable web of interactions between biological, geophysical, and chemical processes and evolution is at the root of a number of conceptual models of soil and landscape evolution. These include Brimhall and others' (1991) invasion–dilation model of soil formation. This emphasizes the importance of channels and burrows in allowing the invasion of material into the soil, after the fauna and flora themselves have promoted upward mass fluxes. Consecutive phases of faunalturbation and root growth constitute an invasive dilational process. Below the depth of roots and burrows, mineral dissolution by descending organic acids leads to internal collapse under the load of the surface soil. The collapsed, softened, condensed horizon is later transformed into soil. Johnson's (1993) dynamic denudation model of landscape evolution also integrates pedologic, hydrologic, biological, and geomorphic processes operating at and between three distinct levels: the weathering front, the subaerial surface, and the stone line or textural contrast at the base of the surficial mantle produced largely by bioturbation (figure 6.4).

Tricart's (1988) more general model integrates soils into ecosystems and explicitly incorporates the coevolution of soils and landforms into a system of mutual interactions between climate, ecosystems, soil, topography, geology, and geomorphic evolution and processes. A key feature is the balance between mass flux and transformation processes operating primarily parallel to the ground surface, which tend to redistribute regolith material, and those acting mostly perpendicular to the ground surface, which tend to create and modify regoliths.

Soils as Complex Nonlinear Dynamical Systems

Spatial variability

If soils are indeed prototype earth surface systems which often behave as complex nonlinear dynamical systems, then complicated,

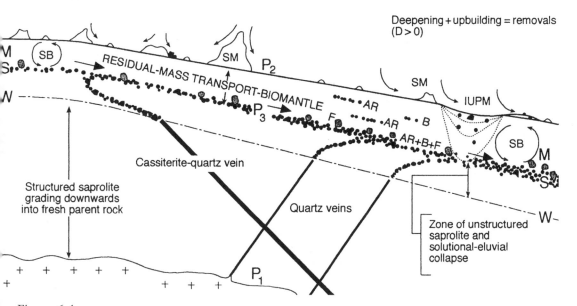

Figure 6.4
An example illustrating Johnson's (1993) dynamic denudation model, in this case in a landscape underlain by dike- and vein-bearing intrusive igneous-metamorphic rock. P_1 is the weathering planation surface, P_2 is the surface erosion planation surface, and P_3 the throughflow planation surface (M, mineral horizon; S, stone line; W, weathered horizon (saprolite); AR, artifacts; B, bioclasts; F, metallic concretions (ferricrete); IUPM, infilling treethrow pit mound; SB, surface bioturbation; SM, surface faunal mounding).

irregular spatial patterns of soil should be common. Indeed, it is the frequent observation of extensive variability of soils and soil properties over small areas and short distances, without observed variations in the soil-forming factors, that has inspired many of the attempts to apply NDS theory to pedology and soil geomorphology (figure 6.5; Culling 1988b; McBratney 1992; Phillips 1993f).

Studies documenting broad ranges of spatial variability and/or extreme short-range variability in specific soil properties at the scale of an individual field or hillslope are common. A few examples are as follows:

- Russo and Bresler (1981) concluded from their measurements of six hydraulic properties in a single field in Israel that their variation could be modeled as though it were a stochastic process. The maximum distance of self-correlation was just 15–55 m.
- Mohanty et al. (1994) measured infiltration at 296 sites on two parallel transects in California. The exponent relating hydraulic conductivity at a given moisture content to the saturated value had a spatial structure that appeared to be mostly random.

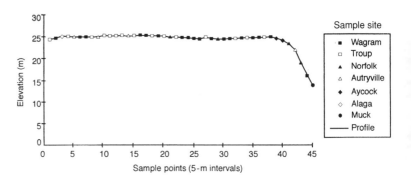

Figure 6.5
*Variation in soil types
(series) along a single
200 m transect near
Falkland, North Carolina
(from Phillips et al. 1994).*

- Anderson and Cassell's (1986) work on physical soil properties in eastern North Carolina indicated, among other things, the number of measurements necessary to estimate the mean of an A-horizon property in a single mapped soil delineation to within ±10 percent. These were 4, 46, and 69 measurements for percentage sand, silt, and clay; 10 for bulk density; 9–35 for soil moisture; and 6 to 1,985 for hydraulic conductivity at various pressures.
- Even among morphologically matched pedons and paired horizons chosen specifically for their similarity, coefficients of variation for percentage sand, silt, and clay, and for moisture content at 1,500 kPa moisture, range from 9 to 40 percent for loess, 23–35 percent for glacial drift, 33–47 percent for alluvium and residuum, 18–32 percent for A- and B-horizons, and 33–51 percent for C-horizons. Coefficients of variation for chemical properties range from 4 to 132 percent (Mausbach et al. 1980).
- Extensive local variability in soil chemistry (pH, Ca, Mg, and N content, and litter mass) was documented by Boettcher and Kalisz (1990) at 135 forest sites in eastern Kentucky.
- The pH of podzolized soils in Iping Common, Britain, shows a highly erratic spatial pattern over 210 m, with a fractal dimensionof 1.7–1.9 (Culling 1986).
- Oliver and Webster (1986a) showed that many soil properties are so variable that they can be treated as spatially dependent random variables. Their examples include pure nugget variance (no spatial dependence at the scale of measurement) for stone and sand content, and unbounded variograms (unlimited variance as distance increases) for pH in Wyre Forest, England. Both nugget and unbounded components characterized the variogram of a gilgai land surface in Wales.

Extensive local, short-range variability is also common with respect to soil types, gross profile morphology, and soil development, as opposed to specific chemical or physical parameters. In southwestern Virginia, Edmonds and others (1985) did not observe taxonomic purity at any level of the US soil taxonomic system at more than half of the randomly selected sites with a 7 m diameter. No site contained a single soil type at the series level. Within a 7 m distance, profiles of three different soil orders may occur (apparently) randomly, and four orders may be represented within a single delineated mapping unit (Edmonds et al. 1985). At Cajon Pass, California, only a portion of the dramatic spatial variations in soil morphology and development could be attributed to observable pedologic and geomorphic controls, to the extent that the unexplained variability calls into question the value of chronosequences (Harrison et al. 1990). Van Es et al. (1991) found high spatial variability in a single 2.2 ha field in the North Carolina Piedmont not only in infiltration and sorptivity, but also a 3.3-fold variation in a color index of soil development.

The spatial variability of soils is of great practical significance to managers, and to scientists not explicitly concerned with spatial patterns or complex soil systems (for reviews see Campbell 1979; Burrough 1983; McBratney 1992; Ibanez et al. 1995). Finke et al.'s (1996) analysis of different means for representing spatial uncertainty in simulating soil functions and behaviors illustrates the critical importance of spatial variability for prediction and management. In semi-arid Botswana, for example, soil variability can be related to microtopography, but is disproportionately large relative to microtopographic variations (Miller et al. 1994). Harris et al. (1994) show how this variability poses serious problems for both farmers and researchers, and contend that microtopographic variations need to be accounted for in agricultural management. Soil variability poses more general problems for land-use planning and soil mapping, as Nortcliff's (1978) studies in Norfolk, England, suggest.

There is also some emerging evidence that soil variability changes systematically as a function of spatial scale or resolution, as would be expected in a nonlinear dynamical system which is unstable and chaotic at some scales but stable and nonchaotic at broader and/or more detailed scales. Ibanez et al. (1995) discuss this and provide examples in an explicit NDS context, and Phillips (1997a) provides some field evidence of this kind of behavior. Without reference to NDS concepts, Sutherland et al.'s (1993) spatial statistical analysis of soil nitrogen in Saskatchewan showed random variation within 11 m grids, but more systematic patterns related to biogeochemical processes in surrounding 110 m grids. Conversely, soilscape variability in contrasting glacial terrains in Wisconsin is large at the regional scale but converges rapidly at more detailed scales, but only if data are stratified by geomorphic subregions (Schaetzl 1986a).

Nonlinear Dynamical Soil Systems

Explicit consideration of soils as complex NDS is still relatively new and rare (for exceptions see Haigh 1987; Culling 1988b; Ibanez et al. 1990, 1995; Huggett 1991, 1995; Nahon 1991; Arlinghaus et al. 1992; Phillips 1993a,c,d; 1993f; Ibanez 1994; Phillips et al. 1996). However, this recent work is presaged by earlier work on soil systems and supported by contemporary research demonstrating complex nonlinear behaviors.

Huggett's (1975) soil-landscape model of soil genesis was an important stimulus to the landscape-scale, complex system-oriented modeling of pedogenesis, and built upon the pioneering efforts of Nikiforoff (1959) and Simonson (1959). Smeck et al. (1983) later proposed an approach to modeling soil systems based on thermodynamic entropy. Muhs (1984) showed how intrinsic pedologic thresholds can explain soil instability in the absence of environmental change, and Parton et al. (1987) used a systems interaction model to explain soil organic matter in Great Plains grasslands. Entropy concepts were applied to pedogenesis and paleosol interpretation by Martini and Chesworth (1992).

Meanwhile, pedologists and soil geomorphologists continue to find evidence of complex feedbacks and instability (Tonkin and Basher 1990; Prosser and Roseby 1995), soil variability disproportionately large compared to the factors which initiated it (Miller et al. 1994; Price 1994), and bifurcating pathways of soil evolution (McDonald and Busacca 1990). Chapter 3 contains numerous other examples.

The Upshot

1 Soils are products of the environment, and represent the interactions of the atmo-, litho-, hydro-, and biospheres.
2 In at least some cases, soils can be interpreted in terms of interacting earth surface system components, and may bear evidence of environmental conditions.
3 Soils coevolve with landforms, ecosystems, and other earth surface systems, and their evolution cannot be decoupled from that of the environment as a whole.
4 The soil is thus a good representative of terrestrial earth surface systems in general.
5 Therefore, whatever generalizations we can make about the behavior of soil systems are likely to apply to ESS in general.

7

Deterministic Complexity and Soil Memory

In chapter 3 we saw a good deal of evidence that a great many earth surface systems are unstable, chaotic, and self-organizing. In chapter 4 I made general, formalized arguments that many ESS behave in this way. In chapter 5 the recognition of such complex behaviors and the incorporation of NDS theory to deal with them were shown to be useful in solving problems. In chapter 6 I tried to build a case for soils as a prototype, canonical earth surface system, arguing that what can be said about soil systems can be said about a good many ESS in general. What I am about to say about them is that many exhibit dynamical instability, chaos, and self-organization, and that this is manifested as soil memory.

Our Story so Far: Chaotic Evolution of Coastal Plain Soils

It all started with the digging of a lot of soil pits, auger holes, and probings in the course of studies attempting to infer long-term patterns of sediment redistribution from soil stratigraphy and morphology. I noticed, like many before (as I subsequently learned), that there was astounding variability of soil types over very short distances and very small areas with no observable variations in topography, vegetation, parent material, or anything else. I saw soil pits where several different soil series were exposed in the same pit. I dug auger holes where apparently identical sites a meter apart might vary by 50, 80, or even 100 cm in the depth to the B-horizon. I encountered prominent pedogenic features, such as spodic concretions, that were present in one sample but absent in a soil a step or two away. And so on. Daniels and Gamble (1967) had passed on the wisdom of the coastal plain soil mappers, whose folklore held that if you wanted to show someone the soil you saw when mapping, you must return to within 15 cm of your auger boring. I began to appreciate the problems and significance of soil variability, and also to wonder if getting within 15 cm would always be sufficient!

It began to occur to me that nonlinear dynamical complexity could explain these complex spatial patterns, where they occurred in the absence of observable variation in environmental controls. If pedogenesis is sensitive to initial conditions, I reasoned, then minor variations in parent material or topography could lead to an increasingly complex soil cover over time. If soil systems are sensitive to small perturbations, I speculated, then changes in soils caused by a tree, an ant colony, or a shovel might persist long after the perturbation disappears and grow large compared to the original disturbance. Some numerical modeling of soil systems I considered plausible for my field area showed that instability and chaos was indeed a strong possibility, and that there was field evidence broadly consistent with such behavior (Phillips 1993c, d, f).

The next step was to devise some empirically field-testable hypotheses. The general approach was to develop equation system or box-and-arrow models based on my field observations in the North Carolina coastal plain. These were based on the processes of textural differentiation. It is the presence, absence, type and thickness of A-, E-, and B-horizons that, along with drainage-related features, produce the major differences between upland soil types (series) in eastern North Carolina. One such analysis produced the hypothesis that soils on older surfaces should be more variable, even where all soil-forming factors other than time are constant. This holds up to the age or time scale at which progressive pedogenetic development begins to level off, as indicated in the work of Markewich et al. (1989), among others (for reviews see Birkeland 1984; Gerrard 1992). Nobody has a good handle on this yet, but my guess based on data in Marilyn Wyrick's (1993) MA thesis is that this takes 400,000–500,000 years in eastern North Carolina. I tested the hypothesis of increasing variability by simply comparing the number of soil types, at the series level, along 0.5 km transects at otherwise identical sites on adjacent geomorphic surfaces of different ages (figure 7.1). At the site on the Pamlico marine terrace, about 75,000–80,000 years, only one soil series was found along the entire transect. At the nearby site on the Talbot terrace (212,000–250,000 years), where parent material, climate, biotic influences, and topography were identical, there were at least seven different series. The application of chaos theory to pedology produced a testable hypothesis related to chaotic pedogenesis, and that hypothesis was confirmed. The details are on the record (Phillips 1993a).

Later, I used a similar model to show that field-testable hypotheses about soil-geomorphic processes (not just whether or not pedogenesis is chaotic) could be generated by the NDS approach (Phillips 1996a). In this case, the analysis suggests that whether weathering is limited by reaction rates or weatherable mineral supplies, and whether vertical translocation is significantly inhibited by the development of B-horizons, are the critical processes in

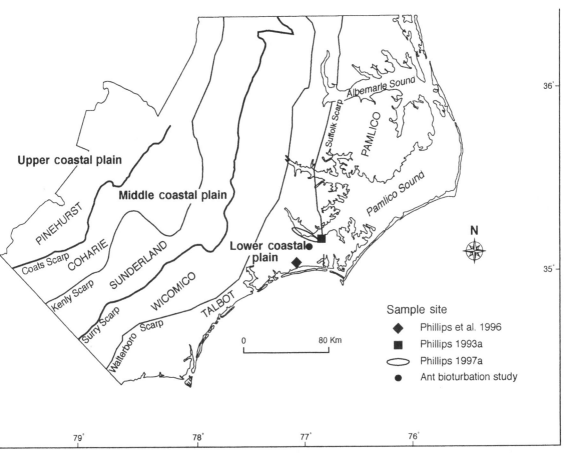

Figure 7.1
The North Carolina coastal plain, showing field sites referred to in the text. The terraces indicated are marine terraces which increase in age, elevation, and degree of dissection from east to west. The scarps are paleoshorelines which are discontinuous on the landscape.

determining whether soils diverge into a complex soil cover or converge toward similar end-states. The controls over weathering processes, and the effects of B-horizons on translocation, are obviously subject to empirical tests.

Finally, I began to look at the issue at a broader scale. As pedogenesis in the region appears to be unstable, chaotic, and self-organizing (if you believe my work, which I did, and do), that means that the variability of soils at the scale of a field or hillslope should show some broader-scale regularity. Chaos, after all, appears random, but the variations occur within definite and well-defined limits. This seemed intuitively to be the case in eastern North Carolina, as despite the well-known variability, there are remarkably consistent soil-landscape correlations at the regional scale. I tested this by comparing soils at three similar sites on the

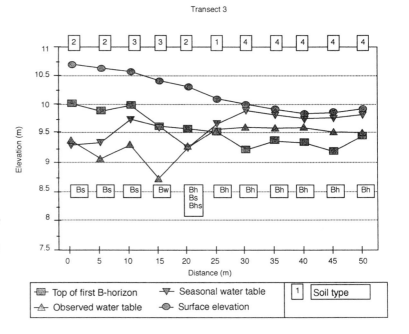

Figure 7.2

Variation in soil types (series) and soil properties along a single transect in Carteret County, North Carolina, from Phillips et al. (1996). Soil types are described in the original source. Bs, Bw, and Bh are types of B-horizons. The seasonal water table is that inferred from soil color and morphology; the observed water table was measured at the times of sampling.

Talbot terrace. My hypothesis was that, on account of unstable and chaotic pedogenesis, there would be a great deal of local-scale variability at each site; that is, many series in a small, homogeneous area. At the same time, there should be general similarities in soils found between the sites. The hypothesis was confirmed: each site had the usual astonishing variety of soil series in a very small area (at least seven series in no more than 0.2 ha), but there was a good deal of consistency and overlap between sites, as at least half the soils found at each site were also found at the other two, and certain related soils occurred at all three sites. Again, the detailed case is on the record (Phillips 1997a).

All the work cited above relates to the upland Ultisols common to the North Carolina coastal plain, so I expanded my work to cover some of the sandy podzolized Spodosols which occur on relict barrier islands (Phillips et al. 1996; figure 7.2). The study site was the Newport Barrier, an area of uniform sandy parent material and age (77,000 years). Soil variability was examined along several transects in terms of soil type, horizon types and sequences, and depths to the B (Bh, Bs, and Bw) horizons. By choosing areas of uniform vegetation and measuring topographic and drainage factors, I was able to control for or account for all known, observable influences on soil variability. The outcome was that correlations between observed controls of soil morphology and soil types and properties were poor, a result one would expect where pedogenesis is chaotic and which indicates either unobservable influences or the long-term persistence of perturbations. A plausible NDS-based explanation rests on the notion that the unstable wetting front and fingered flow phenomena known to exist at the event scale in sandy soils is manifesting itself at broader scales via the restriction of water penetration by B-horizons. Other explanations would have to depend on the persistence of effects no longer observed at the site, which could occur only in an unstable system. The bottom line is that there is field evidence of chaotic pedogenesis, and direct implications with respect to soil-forming processes.

Soil Memory

If pedogenesis is characterized by deterministic uncertainty, then the effects of disturbances or perturbations should be large relative to the magnitude of the perturbation and/or long-lived relative to the time span of the disturbance. In effect, the soil should have a "memory," in that the effects grow unstably and then persist over long periods relative to the perturbation. For example, individual trees are known to influence soil morphology. If there is deterministic uncertainty and soil memory, then those effects will persist long after the tree has rotted way. If not, then the soil morphology will approach the background morphology on a time scale of the same order of magnitude as those of the tree effects, or faster. Unfortunately for paleopedologists and other geoscientists, the memory of individual events or effects may often be incoherent, due to the number of memories present and their overprinting or overlapping in space and time.

At least some pedogenetic events or disturbances are preserved in soils, and the presence of some level of soil memory is fundamental to the study of paleosols. But though it may be clear that persistence of relatively small effects can occur, most scientists need to be convinced that it is common and perhaps pervasive.

Treethrow

Treethrow (also referred to as windthrow and otherwise; see Schaetzl et al. 1989 for a review) occurs when wind (or occasionally some other agent) topples a tree, uprooting the base of the tree and a mass of soil (plate 7.1). In addition to the immediate effects on soil, the depression created by the uprooting, coupled with erosion of the uprooted soil mass and decomposition or combustion of the downed tree, create a characteristic mound–pit topography. Several lines of evidence, and studies from a number of forested locations, suggest that treethrow is quite common (Schaetzl et al. 1989; Vasenev and Targul'yan 1995).

In a review of 21 studies, mostly from North America but also including sites in Europe, Asia, and New Zealand, Schaetzl and others (1989) showed that treethrow mound–pit systems, while covering only a small portion of the forest surface in some areas, typically occupied 10–50 percent of the forest soil surface. Their

Plate 7.1
Treethrow on Big Walker Mountain, Virginia, USA.

review of the literature also suggested that nearly complete canopy destruction by large-scale events such as hurricanes occurs with an average frequency ranging from 100 to 25,000 years, but with a typical frequency in North American forests in the order of 1,000–2,000 years. Data from several locations in European and Asian Russia suggest a periodicity of mass treethrow of 630–1,000 years, while treethrow involving individual trees or small groups at a given forest location has an estimated periodicity of 150–560 years (Vasenev and Targul'yan 1995). Brewer and Merritt (1978) found that a soil pedon in Michigan could be expected to be uprooted at least once every 3,571–5,000 years, and that the entire forest floor is ultimately subjected to treethrow. Scatena and Lugo (1995) showed that 9.4–24.5 percent (varying by geomorphic setting) of a forest surface in Puerto Rico is occupied by tree-uprooting features and another 1.4–14 percent by broken-off tree boles. The estimated forest turnover period due to individual treefall gaps ranged from 60 years in stream valleys to 350 years on mountain slopes (mean for all geomorphic environments 165 years), and the turnover period due to hurricanes was 57 years for all environments (Scatena and Lugo 1995).

Treethrows influence soil morphology via both direct and indirect influences of the resultant microtopography, essentially involving a spatial redistribution of sediment, organic matter, and water. These in turn influence weathering rates, soil chemistry, and translocation (Schaetzl 1990; Schaetzl et al. 1990; Vasenev and Targul'yan 1995). The uprooting and subsequent slumping of soil also results in the mixing of soil horizons, "folding" of pre-existing layers, and sometimes complete inversion of soil profiles (Stephens 1956; Schaetzl 1986b, 1990; Schaetzl et al. 1990). The organic matter supplied by the decomposing tree, as well as the indirect effects via forest and soil organic matter turnover, also influence soil organic properties, chemistry, and weathering (Schaetzl et al. 1990; Scatena and Lugo 1995; Vasenev and Targul'yan 1995). Uprooting may bring gravel and other coarse clasts to the surface, which not only modifies the soil profile but has subsequent effects via surface armoring (Schaetzl and Follmer 1990; Small et al. 1990). Microclimatic effects are also important, as snow accumulation is greater in pits than on mounds. This leads (due to insulation efects) to higher soil temperatures in the pits and increased frequency and duration of impermable ice layers on the mounds (Schaetzl 1990; Schaetzl and Follmer 1990; Vasenev and Targul'yan 1995).

Given that treethrow is common and its effects on soil morphology significant, do these effects persist? In their studies of taiga soils in Kalinin province, the Carpathians, and the Ural mountains, Russia, Vasenev and Targul'yan (1995) questioned the prevailing Russian view that soils disturbed by treethrow return to the background soil. Indeed, the evolution of soils on old treethrows tends toward the background soil, but the question of the longevity of windthrow effects (and their reversibility) is still open. They found prolonged persistence of obvious signs of treethrow in both microtopography and soil-profile morphology which contradict the view that the background soil is restored. Indeed, Vasenev and Targul'yan view the background soil as an attractor, but found that changes in microtopography generally persist up to the next disturbance, and that there are irreversible consequences for soil-profile morphology.

Gravel distributions within treethrow mounds and adjacent undisturbed soils in Wisconsin and Pennsylvania were examined by Small et al. (1990). The erosion of finer material leaves gravels brought to the surface by uprooting as a lag deposit forming surface armors. Continued erosion of remaining nongravelly material diminishes the surface expression of the mound. Gravel armors persist not only long after the tree is gone, but after the mound–pit morphology is no longer evident. The persistence of these treethrow gravel lags in a number of soils was also noted by Johnson (1990).

Schaetzl and Follmer (1990) found that treethrow mounds persist for more than 1,000 years in Michigan and Wisconsin. The longevity is partly due to feedback effects on soils which inhibit erosion. These include gravel armoring as described above, the interception of runoff by upslope pits, and the greater frequency and duration of freezing and impermeable frozen layers on mounds. The persistence, and indeed growth, of soil morphological effects of treethrows on Spodosols in Michigan is associated with varying rates of soil-profile development (Schaetzl 1990). The degree of profile development is greater in apparently undisturbed soils than in pits or mounds, though that of pits may be comparable to undisturbed soils. Rates of progressive pedogenic development, however, are higher in pits and lower on mounds than in the undisturbed soils. This is due mainly to greater moisture, thicker O-horizons, more organic acids, and higher soil temperatures (attributable to greater insulation by litter and snow cover) in the pits, and the greater tendency of impermeable frost layers to develop on the mounds.

In Massachusetts, Stephens (1956) found pit and mounds dating to the early seventeenth and late fifteenth centuries. Even where there was no pit-and-mound topography, evidence of even earlier treethrows was apparent in the traces of overturned soil horizons. On an Ultisol landscape in North Carolina, Phillips and others (1994) found large (>150 cm) and irregular variations in thickness of surficial soil horizons. They argued that the variability, not attributable to any measured or observed variations in topography, parent material, drainage, or contemporary vegetation, was a consequence of the persisting effects of trees, via treethrow and stemflow. The soil stratigraphy of an existing mound–pit system near their study transect showed that subsequent smoothing of the microtopography alone, without any subsequent pedogenetic effects, would result in a 20–60 cm difference in A- and E-horizon thickness over a horizontal distance of less than 2 m. In summary, the pedologic effects of treethrow are a persuasive example of soil memory and deterministic uncertainty.

Ant bioturbation

Colonies of mound-building ants in eastern North Carolina are often concentrated along forest edges, where these localized linear concentrations can have substantial effects on soils. Soil temperature may be the critical control of ant-mound location, but this is speculative. I monitored a 169 m long ecotone between a pine forest and grassy field in Havelock, North Carolina over three years. Surface mounding at a rate of $688\,t\,ha^{-1}y^{-1}$ was found in the 2 m wide edge zone, with negligible mounding in the adjacent forest and rates of less than $0.3\,t\,ha^{-1}y^{-1}$ in the field.

Ants may have numerous physical, chemical, and biological effects on soils. This example focuses on the impacts on surface shear strength: one of the concerns of the monitoring study was the effects of ant mounds on soil erosion. A number of shear vane measurements (at least 25 in each environment) were taken in the grass field within 3 m of the edge zone, and along the edge zone in three different settings: on active ant-hills, on inactive ant-hills from the previous year, and in edge areas without ant-hills, but presumably disturbed by ant-mounding in the past. Measurements were also taken in the forest, but these were very low (mean 0.12 kg cm^{-2}) and not comparable to the mineral soil surfaces due to the well-developed litter/humus layer. Moisture content was 26 percent, but results are presented for comparative purposes only to indicate relative shear strength.

The ant species were not precisely identified. Some, but not all, were *Solenopsis invicta* (fire ant). Others were likely to have been *Formica exsectoides* or *Camponotus abdominalis*, the ant species recognized in eastern North Carolina other than fire ants which are known to build very large mounds. Ant colonies build new mounds every year. Clear evidence of abandoned mounds persists for only a year at the study area, with erosion, settling, and vegetation colonization obscuring evidence of their presence afterward.

Mean shear strength for the grass field was 0.33 kg cm^2 (range 0.15–0.5). By contrast, that of active ant-hills was very low (0.02 kg cm^2; range 0.02–0.04). Settling and compaction in abandoned ant mounds substantially increases shear strength, and the mean for inactive ant-hills was 0.16 kg cm^2. The values clustered around the 0.15–0.18 range, but one anomalous test of 0.75 was recorded. Values for areas of the edge zone without evidence of active or inactive ant mounds ranged from 0.07 to 0.25, with a mean of 0.17. This is essentially the same as that of the abandoned ant mounds. Because measurements were taken in areas of the ecotone without litter and humus layers, the readings are not affected by organic matter. Results thus suggest that changes in surface shear strength associated with the construction and abandonment of ant mounds persist well after the ant colony has gone and evidence of the mound has disappeared.

Root traces

Soil memory of changes and disturbances associated with plant roots is fundamental to the recognition and interpretation of paleosols. The evidence of root effects long after the plant has gone (and sometimes long after the entire ecosystem has gone) results in root traces. Retallack (1990: ch. 3) gives an extensive review.

A full review of the pedologic effects of roots and root channels is well beyond our scope here, but some features associated with roots may persist for long periods. Rhizoconcretions, for example, form because of special local environments created by roots, such as the calcareous rhizoconcretions which may form in dryland soils. Krotovinas form when voids created by decomposition of roots (or animal burrows) are infilled with material differing from the surrounding matrix. Drab-haloed root traces are also common. These are bluish or greenish gray haloes extending into the soil or paleosol matrix as a type of krotovina, or due to anaerobic bacterial activity around the root or root channel which results in chemical reduction. In Retallack's (1994a) review, he cites 15 specific examples of root trace features in paleosols which provide evidence of specific fossil plant formations. This illustrates the long-term persistence of soil changes wrought by short-lived perturbations.

Basket podzols and tree casts

In some soils, particularly podzolized soils, there can be quite dramatic instances of individual tree effects on soils. These are "basket podzols" (plate 7.2) whereby stemflow concentration and organic acids result in leaching of clay, iron, and other materials immediately beneath trees. This results in locally thickened E-horizons and deeper B-horizons, leaving a basket-like effect in soil horizonation. Persistence of podzol baskets after the tree and its stump have gone is an obvious example of soil memory. Such persistence is indicated by the presence of baskets in paleosols and their use as evidence of tree spacing in ancient forests (Retallack 1990: 189). Persistence of locally thickened E-horizons may also be indicated by otherwise unexplained spatial variability of B-horizon depths in Spodosols in coastal North Carolina (Phillips et al. 1996).

Mossa and Schumacher (1993) describe rather compelling evidence for the persistence of pedologic effects of tree roots. Tapered cylindrical pedologic features in South Louisiana soils are shown to be fossil tree root casts (plate 7.3). Trees existing over time scales of decades to a few centuries die, leaving behind a subsurface void as the major tap roots decompose and localized changes in soil chem-

Plate 7.2
Thicker E- and deeper B-horizons associated with a tree that once existed at this site in Croatan, North Carolina, USA. The E–B horizon boundary has been drawn on the photograph.

Plate 7.3
A fossil tree root cast from a Pleistocene paleosol in Croatan, North Carolina, USA. The lighter colored area in the center of the photograph is the former tree trunk, and some preservation of annual growth rings is evident between the spot in the center and the knife on the right-hand side.

istry, bulk density, and porosity. Over time scales of hundreds to thousands of years, microbial activity decreases and sediment is transported into the void. Further changes occur as the feature becomes fossilized, the result being that soil features persist for several orders of magnitude longer than they took to form. The fact that Mossa and Schumacher measured 187 cylinders exposed in five road cuts suggests that the features are widespread, at least in South Louisiana. Mossa and Schumacher (1993) cite other studies where similar features have been examined. I have observed, but not studied, apparent fossil tree casts which are numerous in some Pleistocene paleosols in eastern North Carolina.

Physical disturbances

Many physical disturbances – volcanic eruptions, floods, mining, major erosion events, and so on – undeniably leave long-lasting impacts on soils, but in many cases it is not easy to make the case that these were necessarily minor disturbances whose effects are magnified by instability. However, there are examples that illustrate such tendencies. Sand mining on podzolized soils in Australia disrupted indurated B-horizons, limiting the depth of illuviation. Rapid leaching and deeper illuviation following this short-lived disruption led to thicker E-horizons and deeper B-horizons than existed before the surface sand was removed (Prosser and Roseby 1995). That indirect impacts of sand removal on soil morphology developed and persisted well after the mining ceased illustrates a case of soil memory.

There is also evidence that raindrop detachment and transport in deserts can permanently alter the soil surface. Raindrop impact preferentially removes smaller particles, thus modifying not only the grain size distribution, but also infiltration and microtopography. These feedback effects, according to a simulation model devised by Wainwright et al. (1995), allow desert pavements to form and be maintained. Field tests of the model in Arizona suggest that the raindrop mechanisms in the model only partially predict the formation and maintenance of desert pavements. However, to the extent the raindrop mechanism is operable in forming pavements, it represents a long-term, semi-permanent change in response to a short-lived mechanism, as the raindrop-winnowing no longer operates once the pavement is formed.

Summary

Soils, the canonical example of earth surface systems, may often exhibit instability, chaos, and self-organization. This sort of behavior may once have struck us as exotic but, as we have seen, it is often manifested in a number of rather well-known phenomena. These include the tendency of small changes to persist and grow over time, evolutionary trends toward increasing complexity, and soil memory. One problem, however, is that it is not always obvious when a change is small or large in the context of a particular system or problem, or whether a feature is persisting on a time scale that is long relative to its formative event. That brings us – again – to the problem of scale, and we tackle it in the next chapter.

8

Scale Problems

In everyday commerce a recurrent phrase may be "the check is in the mail," but in the earth and environmental sciences it's more often "it depends on the scale." In this book, spatial and temporal scale has loomed large time and again. In chapter 1's overview of ESS, scale linkage (discussed further below) inevitably cropped up: it is one of the pervasive problems in the geo- and biosciences. In chapter 2, the mathematical background was laid out, and scale emerged as an important implication of the ranges and limits of deterministically complex behavior. In chapter 3, the simultaneous order and complexity of ESS were examined. The importance or observance of one or the other is usually a function of spatial and/ or temporal scale or resolution. This led to the assertion that order and complexity are emergent properties of ESS; that is, one or the other emerges as the dominant mode according to spatial and temporal scales.

In chapter 4, the notion of order and complexity as emergent properties was explored in a more rigorous fashion. We learned that while deterministic complexity is a general feature of ESS, as scale is broadened the irregular, chaotic details become less important and apparent, and broader-scale structures with well-defined limits dominate the view: order emerges from chaos, and is thus an emergent property of the unstable, chaotic system. In the other direction, as scale is narrowed, only a single realization of the system, or a few consecutive realizations, come into view. These are governed by deterministic dynamics, and while the long-term evolution may be inherently unpredictable, the state of the system for the next increment or two is perfectly predictable. Once again, regularity and order emerge from chaos as scale changes.

Specific applications of NDS theory to problem-solving were given in chapter 5, and again scale issues were paramount. The last line of that chapter concludes that the question of spatiotemporal scale, and its relationship to deterministic uncertainty and nonlinear dynamics, is one we must confront. Chapters 6 and 7 addressed soils as a prototype, canonical ESS, and examined relationships between deterministic complexity and specific pedologic phenomena such as large soil changes resulting from small perturbations, and the persistence of soil morphological changes well beyond the time scales of their formative agents. Sometimes these relationships are obvious, or at least decipherable, but the discussion closed with the observation that it is not always obvious when a disturbance is small or large, or whether a feature is persisting on a time scale that is long relative to its formative event. Thus, having repeatedly invoked and encountered scale issues, it is time to confront them head on.

Scale Linkage

Scale linkage – linking processes which operate over fundamentally different time scales – is a critical problem in nearly every geo- and bioscience (Phillips 1988b; Rosswall et al. 1988; Schumm 1988; O'Neill 1989; Huggett 1991; DeBoer 1992; Levin 1992). Attempts to systematically and explicitly cope with scale linkage fall into three broad categories. Hierarchy theory has been proposed as a conceptual framework and pedagogic tool for linking processes at multiple scale-defined hierarchical levels of earth surface systems (O'Neill et al. 1986; Haigh 1987; O'Neill 1988; DeBoer 1992). The utility of hierarchy theory as an operational tool is less convincing than its pedagogic usefulness as a conceptual framework. Second, various mathematical tools have been proposed for translating process descriptions or analyses across spatial scales (for example, Phillips 1988a; King 1991). Third, a number of techniques have been introduced for identifying critical spatial or temporal scales in landscape patterns or process–response relationships (for example, Carpenter and Chaney 1983; Oliver and Webster 1986b; Phillips 1987b).

Scale issues can be lumped into one of four general categories (Delcourt et al. 1983; Gardner et al. 1987; Phillips 1988b; DeBoer 1992; Ehleringer and Field 1993; Blöschl and Sivapalan 1995). First is the problem of pinpointing and measuring the range of spatiotemporal scales of processes, and their characteristic scales. This encompasses not just the operation of process–response relationships, but also the scales of variability and heterogeneity. Second, there are difficulties in associating the characteristic scales of processes with those of observation and of modeling. These issues may boil down to the selection of appropriate spatial or temporal resolutions. The third category of scale problems has to do with operational questions of scale linkage, such as how to upscale from the local to broader scales or downscale in the opposite direction (figure 8.1). Typical approaches involve distribution or interpolation of localized information through space or time, aggregation or lumping, and disaggregation. Fourth, issues of dimensionality

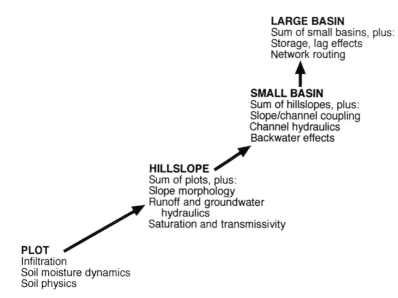

LARGE BASIN
Sum of small basins, plus:
Storage, lag effects
Network routing

SMALL BASIN
Sum of hillslopes, plus:
Slope/channel coupling
Channel hydraulics
Backwater effects

HILLSLOPE
Sum of plots, plus:
Slope morphology
Runoff and groundwater
    hydraulics
Saturation and transmissivity

PLOT
Infiltration
Soil moisture dynamics
Soil physics

Figure 8.1
Scale linkage problems in hydrology. Each level of the hierarchy not only includes the cumulative effects of lower levels, but new considerations as well.

and similarity address the range of scales over which patterns or relationships are constant (or even valid) and rules for up- or downscaling within such ranges.

Here we confront scale linkage at a rudimentary but fundamental level. The goal is to identify scales at which it is important to include particular ESS components and processes and their mutual adjustments versus those at which it is appropriate to consider them independently. For example, given the disparity of time scales over which vegetation and landforms typically exist or evolve, under what circumstances must one explicitly consider their interactions, or treat one or the other as a boundary condition or factor which averages out? Or, given the gap in spatial scales between the mechanics of sediment transport and landscape evolution, to what extent can or should microprocess mechanics be linked to landscape development? At a crude level, these kinds of questions are also one way of dealing with the practical problems of resolution, linkage, and dimensionality/similarity: the answers determine whether the scientist is justified in treating more locally or rapidly operating processes in isolation from broader-scale phenomena, and thus whether the practical problems of scale are relevant.

Much of the recent work on scale has been inspired by problems inherent in dealing with modeling at broad scales of watersheds to the entire earth. These include questions such as how to link processes such as infiltration, which varies at the scale of small patches, with the runoff response of drainage basins, or how to incorporate leaf-scale moisture exchanges into global-scale climate or ecosystem models. Here, scale questions take on particular relevance because of the tendency of ESS to exhibit fundamentally different types of behavior at different scales.

Approaches to Scale Linkage

Landscape sensitivity

Landscape sensitivity is concerned with whether transient or persistent (stable) features – landforms, soils, vegetation communities, and so on – prevail under given circumstances. Brunsden and Thornes (1979) introduced the transient form ratio to address the sensitivity of landforms and landscapes to events causing geomorphic change:

$$TF_r = t_a/t_f \qquad (8.1)$$

Here t_a and t_f are the mean relaxation and recurrence times, respectively. If the transient form ratio exceeds unity, transient forms will prevail, while $TF_r < 1$ indicates the prevalence of characteristic, stable forms (figure 8.2).

It took little imagination to extend Brunsden and Thornes' work to specific considerations of scale linkage with respect to geomorphology and vegetation (Phillips 1995c) and human agency (Phillips 1997b). For example, the recurrence of geomorphically significant ecological changes, t_f, is controlled by the frequency of vegetation disturbances such as fires, and t_a is the time it takes to recover from disturbances. We can also define a transient form ratio for particular ecosystems or plant communities (using the subscript v to indicate vegetation):

$$TF_{r,v} = t_{a,v}/t_{f,v} \qquad (8.2)$$

$t_{a,v}$ is the mean time it takes for succession to restore the pre-disturbance vegetation, and $t_{f,v}$ the recurrence interval of geo-

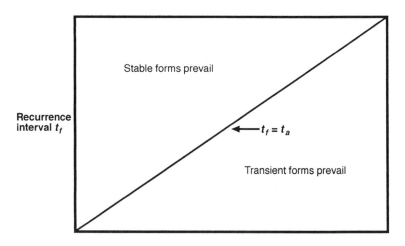

Figure 8.2
The transient form ratio.

morphic disturbances, such as erosional stripping or burial by sediment.

By comparing transient ratios for landforms and vegetation, one can identify domains (cf. Brunsden 1980) where vegetation and geomorphic change are co-dependent (both ratios less than or greater than unity) or at least partially independent (one ratio >1; the other less than unity). Examples using the transient form ratio to determine whether particular phenomena may (or must) be considered independent or co-dependent are given elsewhere (Brunsden and Thornes 1979; Brunsden 1980, 1990; Phillips 1995c, 1997b).

For our purposes the transient form ratio is helpful in the context of landscape memory. The ratio can help in determining whether an ESS is likely to be characterized by disturbance-related forms associated with the persistence and growth of instabilities, or the steady-state forms which may be restored afer the disturbances have faded away. In the treethrow example in chapter 7, t_a is the time it takes following a treethrow event to restore the background soil and t_f is the frequency of tree uprootings. In the Russian Taiga systems of Vasenev and Targul'yan (1995), pedologic effects of treethrow persist for much longer than the recurrence interval of new windfalls. $TF_r > 1$ and the transient, unstable forms of treefall-influenced morphology will prevail, rather than the steady-state form of the background soil.

Transient form ratios can also be used to identify field research questions and critical thresholds. Ryzhova's (1996) analyses, for example, suggest that soil organic matter is sensitive to perturbations, such that even an approximate steady-state balance between litter production and decomposition will be disrupted by any change in net primary production, litter humification, humus mineralization, or import/export of organic carbon. Does soil carbon storage at present represent a good approximation of long-term soil carbon storage (i.e. a stable form), or a transient form reflecting the most recent disturbance? This is a critical question in linking global biogeochemical cycles to climate change. For a given disturbance to soil carbon dynamics such as climate cycles or fire, a frequency could be calculated or estimated. Then the critical relaxation time necessary to achieve the threshold $TF_r = 1$ can be determined, and research can be directed toward determining the relaxation times involved. Conversely, if relaxation times are known then the critical disturbance frequency can be identified. Data from the lower Tar river, North Carolina, for example, suggest that large organic debris is an important energy dissipation mechanism, and that the relaxation time in response to removal of the debris is no more than five years (Phillips and Holder 1991). That is, within five years after debris is removed, new treefalls and accelerated bank erosion will have restored the pre-removal debris regime. Thus, disturbances such as floods or navigational improvements which occur more frequently than twice a decade will result in a prevalence of transient states. Less frequent disturbances mean that a steady-state large organic debris regime will prevail. Such a regime currently prevails in the Tar, but did not during the era of commercial river transport in the eighteenth and late nineteenth centuries, when channel-clearing operations were undertaken every two to five years. An increase in the frequency of floods of sufficient magnitude to clear the channel of organic debris would also result in the transition to an unstable regime.

An approach to dealing with scale linkage that is similar to the transient form ratio is based on comparing relaxation times of ESS with the duration, rather than the frequency, of changes. Chappell (1983) introduced this method in the context of determining whether climate fluctuations represent long-term changes or short-term perturbations to geomorphic systems. I generalized the approach to an examination of state changes in ESS, and whether a given trigger of a state change is best viewed as a singular perturbation or permanent (over a given temporal scale of concern) change in the boundary conditions (Phillips 1995c).

However, while comparisons of relaxation times and durations of endogenous forcings are useful in coping with scale-linkage issues, such comparisons are less useful in dealing with ESS which are likely to be unstable. This is because the unstable growth of small perturbations, rather than basic changes in boundary conditions, may instigate system state changes. Because the geomorphic, stratigraphic, or paleoecological record of state changes in unstable ESS may not reflect the record of changes in boundary conditions, but rather that of singular perturbations, comparing frequencies rather than durations of forcings is more fruitful.

The study of desertification in chapter 5, and the literature on land degradation and range management in general, raise questions about whether particular changes or modes of degradation constitute an unstable, perhaps irreversible state change in the system, as opposed to a transient disturbance to a stable system. The transient form ratio can be employed in this regard, by (for example) comparing the frequency of droughts of an intensity sufficient to result in changes in vegetation communities to the relaxation time for the vegetation to recover from such a drought. In some situations, such as the exposure of calcic horizons, the relaxation time is far longer than any time scales relevant to human land management, and the disturbance may be singular. In this case both t_a and t_f are so long relative to human scales that the transient form ratio is not useful.

Abstracted earth surface systems

What about relationships among components of earth surface systems which vary over different scales? To what extent are they independent or interdependent? Some phenomena operate over such different scales that it is intuitively obvious that they may be treated in isolation. There are surely links between cell biology and ecosystem ecology and between tectonic movements and aeolian saltation. But microbiologists and ecologists, and tectonophysicists and aeolian geomorphologists are generally on safe ground as they pursue their subjects separately. Likewise, other phenomena vary over somewhat different scales but are similar enough so that their interactions must both be taken into consideration. Rainfall-runoff modeling, for example, must consider both the rapid response of infiltration and overland flow and the more delayed response of subsurface flow and runoff routing through channels. Similarly, the time scales of uplift and those of slope erosion and mass wasting episodes may differ, but both must be considered in studies of long-term landscape evolution.

But the situation is not always so clear. ESS are affected by numerous factors acting over a wide range of scales. Sediment entrainment and some forms of mass wasting, for example, are virtually instantaneous. Sea floor spreading and orogenies, conversely, operate over millennial time scales, and there are numerous processes and responses at a number of intermediate time scales. Clearly, in some cases, such as broad-scale tectonics, one can dismiss geologically recent and short-lived influences such as human agency due to the disparity in temporal scales. In other cases, such as urban geomorphology, one can dismiss a number of long-term geological processes for the same reason. But consider the dramatic impacts of humans on many rivers. Are these significant to evolution of the fluvial system? How different must the time scales be to allow human agency to be addressed independently of longer-term or ongoing geological phenomena? This is the sort of question addressed by scale independence in abstracted systems.

Let's start with our classic nonlinear dynamical n-component ESS. Imperfect knowledge and issues of parsimony and tractability suggest that we can rarely include (or even identify?) all n components of the real-world earth surface system. Thus, we construct abstracted, m-component systems, $m < n$:

$$dx_i/dt = f(x_1, x_2, \ldots, x_m; c_1, c_2, \ldots, c_m); i = 1, 2, \ldots, m \qquad (8.3)$$

Equation 8.3 is simply an abstracted version of the classic typical NDS (see chapter 2). As we have seen, an equation system in the form of equation 8.3 can be translated to an $m \times m$ interaction matrix, A(a). A is the $n \times n$ matrix for the full (unknown) system. Matrix A(d) could be formed by deleting rows and columns 1 through m of A (two other submatrices of A would exist, representing the interrelations of the abstracted and ommitted components with each other). Matrix A(d) then reflects the mutual interactions of the components of the system omitted from the abstracted model.

Schaffer (1981) proved the product theorem for abstracted eigenvalues, and even though his paper concerned ecological systems, it set the stage for Phillips' (1986b) formal demonstration of scale-independence arguments such as those of Schumm and Lichty (1965) in geomorphology. Schaffer (1981) obtained the result below for cases where the elements of A(a) are assumed to be of the order of 1 and the elements of A(d) of the order of $>>1$. Order in this case is the scale over which the components vary.

$$\Delta(\lambda) = [\det(A(a) - \lambda I)][\det(A(d) - \lambda I)] \tag{8.4}$$

$\Delta(\lambda)$ is the characteristic polynomial of a matrix similar to A and I is an identity matrix. Similar matrices have identical eigenvalues, and determinants equal the product of the eigenvalues. Equation 8.4 shows that if the abstracted and omitted components operate on very different time scales, the eigenvalues of A can be separated into two separate, independent components. The separation of A(d) and A(a) indicates that ESS phenomena operating at time scales an order of magnitude or more different are effectively independent of each other, despite apparent relationships in the field. Thus grain saltation is independent of tectonic uplift, and studies of either need not account for the other. On the other hand, erosional unloading and uplift may operate over roughly similar temporal scales, indicating that they are not independent.

If one is searching for the scale at which order emerges from deterministic complexity, the abstracted systems argument can help by allowing one to distinguish variables or phenomena likely to be independent of each other due to disparate scales. Local-scale erosion, for example, typically operates at spatial scales of $10^0–10^1$ m. In many cases erosion is characterized by unstable spatial divergence into zones of erosional removal, accretion, and transport (or an approximate balance between erosion and deposition). The scale independence argument suggests that orderly patterns should be observed at scales of $10^2–10^3$ m. This is exactly the type of phenomenon reported by Pickup and colleagues (Pickup and Chewings 1986; Pickup 1988) in the formation of scour–transport–fill sequences in Australia, and Engelen and Venneker (1988) in the formation of similar sequences in Italian rivers.

My field studies in eastern North Carolina suggest that pedogenesis is chaotic at length scales of the order of $10^0–10^2$ m (see chapter 7). The abstracted systems theory then suggests that order should emerge at length scales of the order of $10^3–10^4$ m. This is exactly what I found in an effort to explicitly address this issue (Phillips 1977a), and also corresponds to the broad scale of soil-landscape correlations.

The information criterion

Global climate models commonly rely on the Courant–Friedrichs–Lewy (CFL) criterion, to make sure that differencing schemes are stable (Martin, 1993). The CFL criterion is necessary because of the broad range of spatial and temporal resolutions for various components of climate and surface–atmosphere transfer models. The CFL criterion may be generalized to a broader information criterion. Beyond computational concerns, it is necessary to constrain models and analyses so that changes are not propagated across the landscape faster than they actually occur in time. For example, vegetation changes are virtually continuous and thus usually faster than geomorphic evolution or climate change. The information criterion is useful in ESS models to identify time steps and spatial resolutions so that slowly changing phenomena such as topography or climate can register the results of more rapid phenomena, such as vegetation or human agency (see Martin 1993; Phillips 1995c, 1997b).

The information criterion is useful where the concern is with a rapidly-operating phenomenon (r) interacting with a slower one (s). Then fidelity of scales dictates that

$$\Delta t_r < \Delta s_r / c \quad \text{and} \tag{8.5a}$$

$$\Delta t_s > \Delta s_s / m \tag{8.5b}$$

Δt and Δs are the time steps and the spatial resolution (grid spacing), respectively, c is the rate of the rapid phenomenon, and m is the rate of the slow one.
We then obtain

$$\Delta s_s / \Delta s_r < (m/c)(\Delta t_s / \Delta t_r) \quad \text{and} \tag{8.6a}$$

$$\Delta t_s / \Delta t_r > (c/m)(\Delta s_s / \Delta s_r) \tag{8.6b}$$

If $\Delta s_s = \Delta s_r$ owing to interest in a particular landscape unit, then the temporal resolution should be such that the time steps for the slow process are greater than those for the faster one by the ratio c/m. If the time steps are fixed and equal, the slower component should be examined at a spatial resolution more detailed than that of the faster one by the ratio m/c. If both the spatial and temporal resolutions are fixed, for example in a cellular model with an annual time step, equations 8.6a and b provide a means for evaluating whether it is appropriate to include both components within the analysis.

For example, two studies already cited have documented extreme local spatial variability in soil surface horizon thickness in the North Carolina coastal plain. One provides evidence of deterministic uncertainty (Phillips et al. 1996), and both (Phillips et al. 1994, 1996) suggest that the effects of individual trees on the depth of B-horizons have left lasting imprints on soils. One way to explore that question further would be via simulation models of the coevolution of vegetation and soils.

The development of soil and vegetation could perhaps be modeled on a plot-by-plot basis. What, then, would be the appropriate time steps? In this case spatial resolution is fixed, and in equation 8.6b $\Delta s_s / \Delta s_r = 1$. Then the pedologic time step should exceed the vegetation time step by the ratio c/m. Well-developed Ultisols can develop on the North Carolina Coastal plain in less than 212,000 years, as indicated by their presence on surfaces of that age (Daniels et al. 1978). Well-developed Spodosols are found in 77,000 year relict barrier sands (Phillips et al. 1996). If the vegetation time step is one year, and mature forests develop in 200 years, then the pedologic time step should be about 1,125 years or more for Ultisols and 385 for Spodosols.

Alternatively, we could try to use chronosequences, historical data, or soil stratigraphy to examine the coevolution of soil and vegetation over a given and fixed time interval (i.e. $\Delta t_s / \Delta t_r = 1$). Then, equation 8.6a suggests that pedogenesis would need to be assessed at a spatial scale more detailed than that of vegetation by the ratio m/c. Using the same numbers as above, this implies that if soils are examined on a pedon-by-pedon basis with a surface area of $1\,m^2$, forest change should be evaluated over plots more than $1.1\,ha$ in area for Ultisols, or about $0.4\,ha$ for Spodosols.

Rates and Relationships

The problem of determining spatiotemporal scale relationships is tied closely to an understanding of the rates, durations, and frequencies of processes and formative events in earth surface systems, and to an understanding of relaxation times in response to disturbances. To the extent these phenomena are understood, the techniques described above – and other, similar tools – can be applied. Then the earth system scientist can move beyond vague notions of changes or perturbations, long-term or short-term persistence, and other such notions to define, or at least reliably estimate, exactly what constitutes a perturbation (as opposed to an ongoing, intermittent forcing) or long-term persistence.

This is important in bringing the nonlinear dynamics perspective into line with more reductionist approaches to earth surface systems. The systems analyst cannot afford, ultimately, to deal strictly with abstract systems, numerical simulations, or amenable time series data. The question of scale is crucial to an understanding of complex system dynamics, and every path to such an understanding is paved with knowledge of real-world rates, frequencies, relaxation times, and spatial dimensions. There is no magic nonlinear formula or complex systems elixir for understanding ESS. We must still go into the field and measure or infer how far, how fast, how many, and how much.

9

Walk that Aisle

Brother, you gotta walk that aisle!
Nature Boy Ric Flair

In the morality plays that are professional wrestling, whence Ric Flair's comment arises, sooner or later the showdown is at hand, and antagonists (or protagonists) must indeed "walk that aisle" to the moment of truth in the squared circle. Now the time has come for me to walk that aisle; to put forth my version, or vision, of what the preceding chapters mean. The goal was, and is, to move toward a theory of earth surface system behavior by developing some principles of how they work. There's the bell . . .

Principles of Earth Surface Systems

Principle 1: Earth surface systems are inherently unstable, chaotic, and self-organizing
A review of the empirical evidence shows that deterministic complexity occurs in many ESS (chapters 3 and 5). Field evidence shows that soils, a prototype ESS, exhibit deterministic complexity (chapter 7). Highly generalized, canonical models applicable to a variety of ESS are unstable and chaotic (chapter 4). Most importantly, systems that are characterized by divergence or differentiation over time, or by known but unmeasured (or unmeasurable) underlying controls, must exhibit deterministic complexity (chapter 4). Virtually all ESS share the latter traits; otherwise we would have homogeneous landscapes and a clear view of the factors controlling spatial variability.

This is not – repeat, not – to say that all ESS are unstable all the time. They may be so only under certain conditions or at certain times. Those conditions and times may not be relevant to some questions and problems. Even more emphatically, this is not to say that all ESS are unstable at all scales – in fact, quite the opposite, as stated in principle 2 (below). Again, the scales at which deterministically complex behavior occurs may be unimportant or uninteresting in many contexts.

Principle 2: Earth surface systems are inherently orderly
The dynamical instability and chaos of ESS, and their link to self-organization, make order in ESS inevitable. Deterministic chaos in a nonlinear dynamical system is governed by an attractor. Accordingly, the system occupies a limited portion of its state space; not all states or combinations of variables or components are possible, and not all are equally likely. This means that such a system is decidedly not random. Further, instability is finite, chaos occurs within definite limits, and self-organization is self-organization. If you start with a deterministically complex pattern and broaden the scale, you must eventually reach a point where the finite limits and organized patterns become dominant, rather than the pseudo-random weirdness within them. Thus chaotic pedogenesis and wild spatial variability at the local scale resolves itself into the comfortable regularities of soil-landscape relationships; the chaotic turbulence of fluids doesn't prevent water from flowing downhill or wind from blowing according to pressure gradients; and the chaotic evolution of channel networks is ultimately manifested in the familiar topologies that converge toward similar patterns worldwide. And so on.

In the other direction, while chaotic systems are unpredictable in detail in the long run, as you sharpen the detail, they are perfectly predictable a few iterations into the future. Likewise, as the resolution is increased and the scope decreased, complex interactions that lead to the unstable growth of perturbations disappear, as components involved in those interactions disappear or become irrelevant. So the chaotic interactions of fluid molecules become tractable if you get it down to just a few particles. And the overlapping, persisting "memories" that soils have of events or perturbations can be disentangled and understood when the focus is on a single, contemporary event. And the pseudo-random branching growth of channel networks becomes amenable to process mechanics when a single channel in a single time and place is contemplated. And so on.

Principle 3: Order and complexity are emergent properties of earth surface systems
That is, both orderly, regularly, stable, nonchaotic patterns and behaviors and irregular, unstable, chaotic ones appear and disappear as spatial and/or temporal scale or resolution is increased or decreased. This follows from principles 1 and 2. In Straub's (1996) debris flow example, interparticle collisions in highly sheared granular flow are governed by deterministic chaos, making them sensitive to initial conditions and unpredictable. However, as one

increases the resolution to individual particles, one can readily predict behavior with just a few simple equations. And as one expands to the bulk behavior of granular flows, order emerges in the form of a simple linear relationship between kinetic energy (drop height) and travel length.

Principle 4: Earth surface systems have both self-organizing and non-self-organizing modes
This follows in part from the first three principles. It is important, however, to note that some ESS may also operate in either mode at the same scale. Topographic evolution, for example, may be either self-organizing with increasing relief or non-self-organizing with decreasing relief. It is also worth reiterating that neither stable/nonchaotic/non-self-organizing nor deterministically complex behavior is inherently more common, important, or "normal" than the other.

Principle 5: Both unstable/chaotic and stable/nonchaotic features may coexist in the same landscape at the same time
As many ESS may exhibit either mode, different locations in the same landscape may, at a given time, exhibit convergent or divergent developmental trends. This is evident in the concept of "complex response" in geomorphology, where different parts of the same system may respond differently at a given time to the same stimulus (Schumm 1988, 1991); for example, when channel incision in tributary headwaters results in valley aggradation in trunk streams. Further, as different landforms and ecosystems may be controlled by factors operating over a range of spatial and temporal scales, these multi-scale controls may result in both stable and unstable features in the same landscape.

Principle 6: Simultaneous order and regularity may be explained by a view of earth surface systems as complex nonlinear dynamical systems
Both order and disorder are frequently observed in real landscapes and processes. This can arise from stochastic forcings and environmental heterogeneity superimposed on more orderly, regular patterns. However, some combination of both order and complexity is inevitable in ESS regardless of the extent to which stochastic forcings and environmental heterogeneity may be involved.

Principle 7: The tendency of small perturbations to persist and grow over finite times and spaces is an inevitable outcome of earth surface system dynamics
The growth of small changes is well known to scientists. Thus the enlargement of nivation hollows, the development of bedforms, and the deepening of weathering depressions, to give a few examples, are particular cases of a general phenomenon. Obviously

not all minor perturbations persist or grow. An understanding of the stability properties of nonlinear dynamical ESS can help us determine which perturbations will grow under what circumstances.

Principle 8: Earth surface systems do not necessarily evolve toward increasing complexity

This arises logically from the preceding principles, particularly principle 4. However, because any consideration of evolution is likely to be informed and indirectly influenced by concepts of biological evolution, this is worth stating as a separate item. Biological evolution on earth, at least so far, has been characterized by increasing biodiversity and increasing complexity of organisms. This does not translate to the ecosystem scale, however, or to earth surface systems in general. ESS may get more or less complex at any given scale, and may do either at a given time.

Principle 9: Neither stable, non-self-organizing nor unstable, self-organizing evolutionary pathways can continue indefinitely in earth surface systems

Unstable, chaotic, self-organizing trends are inherently self-limiting, and in some situations stable pathways are too. In any event, stable development implies convergence, which must ultimately lead to a lack of differentiation in time or space. The maintenance of such a stable static condition not only seems unlikely given the evidence at hand, but would require an absence of disturbances large enough to reconfigure the system. The earth's history does not suggest that this ever has been, or will be, the case, particularly over geological time scales.

Principle 10: Environmental processes and controls operating at distinctly different spatial and temporal scales are independent

This principle could be deduced as an implication of stability and instability as emergent properties. It can also be derived independently based on the theory of abstracted earth surface systems (see chapter 8).

Principle 11: Scale independence is a function of the relative rates, frequencies, and durations of earth surface phenomena

Tools such as the transient form ratio and abstracted ESS theory can be used to determine whether the scales over which phenomena operate are different enough to be independent. Tools such as the information criterion can be used to determine appropriate resolutions for linking phenomena which operate at disparate scales. In all cases the necessary information involves process rates, the frequency of formative events, the duration of responses, or relaxation times. Both historical perspective and measurements of process rates are necessary to address scale linkage, scale independence, and (in)stability as an emergent property.

The (D)Evolution of Diversity

Do earth surface systems become more diverse over time? As we have seen, the answer may be yes, no, or maybe, depending on the spatial and/or temporal scale, and upon historical contingency; that is, the dynamical mode an ESS happens to be in, or its history and memory of disturbances. A related question, of crucial importance to both science and natural resource management, is whether the diversity of various components of ESS change in the same direction.

Over very long, geological time scales, the answer seems to be that changes in bio-, pedo-, topo-, litho-, hydro-, and climodiversity are mutually reinforcing. Retallack (1990) marshals a great deal of paleontological evidence to show that organisms and soils have coevolved since the Cambrian toward greater species richness and soil diversity (though at varying rates and with interruptions and short-term reversals). The pedodiversity is also influenced by lithology, climate, topography, and other factors, thereby raising the possibility of increasing diversity of these factors as well. Huggett's (1991) examination of climate change and earth history concludes that climate, biota, landforms, and soils coevolve in mutually reinforcing ways, and that many of the changes may be driven by the internal nonlinear dynamics of the mutual interactions.

With respect to biodiversity, the overall long-term trend in biological evolution is toward increasing richness and complexity, albeit punctuated by episodes of extinction and extirpation. In the spatial domain, Rosenzweig (1995) argues forcefully that area is the primary control over species diversity, and attributes this largely to the increased availability of different habitats (diversity in climate, hydrology, soils, topography, and so on). The latter clearly plays a major role, and Scheiner and Rey-Benayas's (1994) study of global-scale plant diversity shows that, in addition to larger areas and warmer temperatures, the most diverse landscapes are associated with higher relief (topodiversity) and large seasonal temperature changes (climodiversity). Ibanez (1994) documents the coevolution of topodiversity (via fluvial dissection), pedodiversity (due to topographic, parent material, and topographically controlled microclimate influences on soil), and biodiversity over Quaternary time scales in the Iberian peninsula. Ibanez et al. (1995) further show that pedodiversity tends to increase with area, due to the increased variation in topography, local or microclimate, geological influences, and biogeography.

The relationships between bio-, pedo-, topo-, hydro-, and climodiversity (lithological diversity is considered to be captured by topography and soils) are shown in figure 9.1. The feedback relationships are all positive, indicating mutual reinforcements. Diversity in climate, soils, topography, and hydrologic regimes creates variations in habitats and niches which lead to greater biodiversity; greater homogeneity in these habitat factors inhibits biodiversity. Likewise, the degree of variation in these factors of soil formation

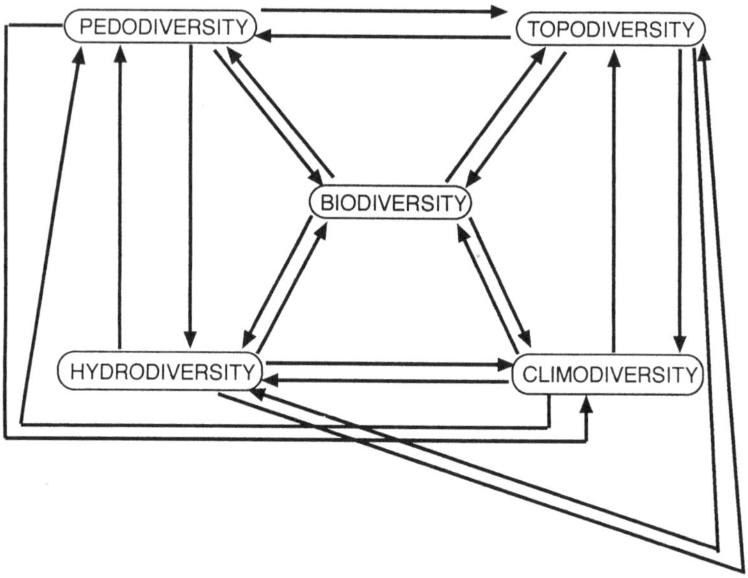

Figure 9.1
The relationships between bio-, pedo-, topo-, hydro-, and climodiversity. All links are positive. For a particular problem certain links may be missing or unimportant, but all plausible configurations involve only positive links.

is directly, albeit nonlinearly, related to the degree of soil diversity. Hydrologic regimes are dependent upon climatic moisture inputs, hydraulic and hydrologic properties of soils, moisture use, exchanges, and concentration by organisms, and by topographically influenced moisture fluxes. Thus hydrodiversity must be enhanced or inhibited by increased or decreased diversity of the latter factors. Topographic change due to weathering, erosion, and deposition is clearly influenced by biota, soils, hydrologic regimes, and climate. At particular temporal or spatial scales one or more of those factors might be negligible (an arrow would be removed from figure 9.1), but there are no plausible circumstances where increased diversity of biota, soils, hydrology, or climate would lead to *less* diverse topography or rates of topographic change, and vice versa. The situation is analogous for climate: particular effects of topography, moisture fluxes and storage, soil, and organisms may not be important at some spatial or temporal scales, but in any case the relationships that are operable would be positive.

One could, for a given situation, eliminate arrows from figure 9.1. But in any case you would have a diagram with all positive arrows, and an interaction matrix with all positive entries. Such a system must be unstable, indicating that a change in any component could cause the entire system state to change. In fact, the history of the earth is replete with such evidence. Extraterrestrial impacts, for example, have set off unstably growing perturbations which altered the entire state of the earth system (Huggett 1990; Shaw 1994). Changes in topodiversity associated with paleogeographical changes as the continents drifted led to global changes in climate, runoff, and weathering (Huggett 1991; Otto-Bleisner 1995).

The implications for management, necessarily at much more

restricted scales, are too numerous to discuss even superficially, but the overriding theme must be obvious: increases or decreases in the diversity of any environmental component are likely to be accompanied by similar changes in the other components. This confirms the practice of protecting biodiversity by protecting and enhancing habitat diversity. It also implies that the loss of biodiversity will have broader environmental repercussions.

At geological and evolutionary time scales, the implication is that there must be fundamental limits to environmental diversity and complexity. First, this follows from the principles above: the system (figure 9.1) is unstable and chaotic, so the growth of perturbations must be asymptotically limited. Second, this follows from the inherent limits on the growth of some environmental components. Topodiversity cannot increase (or decrease) indefinitely, as there are fundamental geophysical limits on the height of mountains, the depth of incision or planation, and the steepness or flatness of slopes (Ahnert 1984; Phillips 1995b; Miller and Dunne 1996). Climate and hydrologic changes also apparently cannot continue indefinitely along any given trend before limits and negative feedbacks kick in (Huggett 1991). Because bio- and pedodiversity seem to be clearly controlled by area (Huggett 1995; Ibanez et al. 1995; Rosenzweig 1995), this suggests upper limits. The argument that increasing biological complexity can occur due to strictly biological mechanisms is rejected, due to the obvious influence exerted by geophysical and geochemical factors, and by arguments such as those of Brooks and Wiley (1988) who assert that natural selection and other proximal factors primarily determine the rates of evolution, not the direction.

Even at the broadest scales, earth surface systems – indeed, the earth system – cannot be viewed as proceeding along a particular developmental pathway, either toward ever-increasing diversity or toward any stable end-state. There are multiple possible pathways and many possible destinations. Prediction does not depend on interpreting earth history as an inevitable, predetermined sequence of events. Rather, prediction depends on interpreting earth history as a sequence of historically contingent events and system states from among a much larger population of possible states. By determining the circumstances under which particular pathways were followed, or under which particular states emerged, we can move toward a more complete understanding of our planet and our relationship to it.

Walk by Faith

Scientists, particularly in the empirical field sciences, are confronted with a beautiful and fascinating, but complex and confusing world. At times this can make our meager efforts to understand it seem incredibly, overwhelmingly difficult – even impossible. The reductionists among us have long taken psychological refuge in the view that if only we can get more and better measurements we will slowly and surely disentangle the mess. Accordingly, the notion that there is inherent, irreducible complexity that cannot be understood by reductionist means can have a disquieting effect. Some have been known to respond, in either defeat or defiance, by suggesting that if the world is so damn chaotic, then we should all just go home and have a beer.

Let us by all means go home and have a beer. But first, recall from chapter 1 that reductionist studies are a necessary precursor to, and companion of, systems-oriented approaches. More fundamentally, there is also a form of psychological refuge available from the perspective of complex nonlinear dynamics. This is that the uncertainty, irregularity, and complexity that is not amenable to reductionism must inevitably be part, at some scale, of a broader pattern of order, regularity, and self-organization. Thus, like the reductionist who takes comfort in the ideal that understanding will come if only we can know the details that we just can't get at yet, I take comfort from the view that the indecipherable weirdness is part of some broader pattern that we just haven't recognized yet.

Thus reassured, we walk that aisle. As ever, we walk by faith and not by sight.

The principles of earth surface systems

---

1 Earth surface systems are inherently unstable, chaotic, and self-organizing
2 Earth surface systems are inherently orderly
3 Order and complexity are emergent properties of earth surface systems
4 Earth surface systems have both self-organizing and non-self-organizing modes
5 Both unstable/chaotic and stable/nonchaotic features may coexist in the same landscape at the same time
6 Simultaneous order and regularity may be explained by a view of earth surface systems as complex nonlinear dynamical systems
7 The tendency of small perturbations to persist and grow over finite times and spaces is an inevitable outcome of earth surface system dynamics
8 Earth surface systems do not necessarily evolve toward increasing complexity
9 Neither stable, non-self-organizing nor unstable, self-organizing evolutionary pathways can continue indefinitely in earth surface systems
10 Environmental processes and controls operating at distinctly different spatial and temporal scales are independent
11 Scale independence is a function of the relative rates, frequencies, and durations of earth surface phenomena

---

References

Abrahams, A. D. 1984. Channel networks: a geomorphological perspective. *Water Resources Research* 20: 161–88.

Abrahams, A. D., Parsons, A. J., and Wainwright, J. 1995. Effects of vegetation change on interrill runoff and erosion, Walnut Gulch, Arizona. In C. R. Hupp, W. Osterkamp and A. Howard (eds), *Biogeomorphology, Terrestrial and Freshwater Systems*. Proceedings of the 26th Binghamton Geomorphology Symposium. Amsterdam: Elsevier, pp. 37–48.

Afouda, A. 1989. Consistent scale parameter for hydrological studies. In *FRENDS in Hydrology*. Wallingford, UK: International Association of Hydrological Sciences Publication 187: 107–18.

Ahnert, F. 1967. The role of the equilibrium concept in the interpretation of landforms of fluvial erosion and deposition. In P. Macar (ed.), *L'Evolution des Versants*. Liege, France: University of Liege Press, pp. 23–41.

Ahnert, F. 1976. Brief description of a comprehensive three-dimensional model of landform development. *Zeitschrift für Geomorphologie suppl.* 35: 1–10.

Ahnert, F. 1984. Local relief and the height limits of mountain ranges. *American Journal of Science* 284: 1035–55.

Ahnert, F. 1987. Process-response models of denudation at different spatial scales. *Gatena suppl.* 10: 31–50.

Ahnert, F. 1988. Modelling landform change. In M. G. Anderson, (ed.), *Modelling Geomorphological Systems*. Chichester: John Wiley, pp. 375–400.

Ahnert, F. 1994. Modelling the development of non-periglacial sorted nets. *Catena* 23: 43–63.

Allen, J. C. 1990. Factors contributing to chaos in population feedback models. *Ecological Modelling* 51: 281–98.

Allen, J. R. 1974. Empirical models of longshore currents. *Geografiska Annaler A* 56: 237–40.

Amundson, R. and Jenny, H. 1991. The place of humans in the state factor theory of ecosystems and their soils. *Soil Science* 151: 99–109.

Anderson, S. H. and Cassell, D. K. 1986. Statistical and autoregressive analysis of soil physical properties of Portsmouth sandy loam. *Soil Science Society of America Journal* 50: 1096–104.

Arlinghaus, S. L., Nystuen, J. D., and Woldenberg, M. J. 1992. An application of graphical analysis to semidesert soils. *Geographical Review* 82: 244–52.

Armstrong, A. C. 1980. Soils and slopes in a humid temperate environment: a simulation study. *Catena* 7: 327–38.

Baker, W. L. and Walford, G. M. 1995. Multiple stable states and models of riparian vegetation succession on the Animas River, Colorado. *Annals of the Association of American Geographers* 85: 320–38.

Balling, R. C. Jr 1988. The climatic impact of a Sonoran vegetation discontinuity. *Climatic Change* 13: 99–109.

Balling, R. C. Jr 1989. The impact of summer rainfall on the temperature gradient along the United States–Mexico border. *Journal of Applied Meteorology* 28: 304–8.

Barrett, L. R. and Schaetzl, R. J. 1993. Soil development and spatial variability on geomorphic surfaces of different age. *Physical Geography* 14: 39–55.

Barrow, C. J. 1991. *Land Degradation.* Cambridge: Cambridge University Press.

Barth, H. K. 1982. Accelerated erosion of fossil dunes in the Gourma region (Mali) as a manifestation of desertification. *Catena* suppl. 1: 211–19.

Beatty, S. W. 1987. Origin and role of soil variability in southern California chaparral. *Physical Geography* 8: 1–17.

Beer, T. and Borgas, M. 1993. Horton's laws and the fractal nature of streams. *Water Resources Research* 29: 1475–87.

Bennett, K. D. 1993. Holocene forest dynamics with respect to southern Ontario. *Review of Palaeobotany and Palynology* 79: 69–81.

Beven, K. 1996. Equifinality and uncertainty in geomorphological modelling. In B. Rhoads and C. Thorn (eds), *The Scientific Nature of Geomorphology*, Proceedings of the 27th Binghamton Geomorphology Symposium. New York: John Wiley, pp. 289–313.

Bhattacharya, K. 1993. The climate attractor. *Proceedings of the Indian Academy of Sciences (Earth and Planetary Sciences)* 102: 113–20.

Birchfield, G. E. and Ghil, M. 1993. Climate evolution in the Pliocene and Pleistocene from marine-sediment records and simulations: internal variability versus orbital forcing. *Journal of Geophysical Research* 98B: 10385–400.

Birkeland, P. W. 1984. *Soils and Geomorphology.* Oxford: Oxford University Press.

Bishop, P., Young, R. W. and McDougall, I. 1985. Stream profile change and longterm landscape evolution: early Miocene and modern rivers of the east Australian highland crest, central New South Wales, Australia. *Journal of Geology* 93: 455–74.

Blöschl, G. and Sivapalan, M. 1995. Scale issues in hydrological modelling: a review: *Hydrological Processes* 9: 251–90.

Boettcher, S. E. and Kalisz, P. J. 1990. Single-tree influences on soil properties in the mountains of eastern Kentucky. *Ecology* 71: 1365–72.

Boston, K. G. 1983. The development of salt pans on tidal marshes, with particular reference to south-eastern Australia. *Journal of Biogeography* 10: 1–10.

Bourke, M. C. 1994. Cyclical construction and destruction of flood dominated flood plains in semiarid Australia. In *Variability in Stream Erosion and Sediment Transport*. Wallingford, UK: International Association of Hydrological Sciences Publication 224: 113–23.

Brewer, R. and Merritt, P. G. 1978. Wind throw and tree replacement in a climax beech–maple forest. *Oikos* 30: 149–52.

Brimhall, G. H., Chadwick, O. A., Lewis, C. J., Compston, W., Williams, I. S., Danti, K. J., Dietrich, W. E., Power, M. E., Hendricks, D., and Bratt, J. 1991. Deformational mass transport and invasive processes in soil evolution. *Science* 255: 695–702.

Brinkmann, W. L. F. 1989. System propulsion of an Amazonia lowland forest: an outline. *GeoJournal* 19: 369–80.

Brooks, D. R. and Wiley, E. O. 1988. *Evolution as Entropy*, 2nd edn. Chicago: University of Chicago Press.

Brooks, S. M. and Richards, K. S. 1993. Establishing the role of pedogenesis in changing soil hydraulic properties. *Earth Surface Processes and Landforms* 18: 573–88.

Brunsden, D. 1980. Applicable models of long term landform evolution. *Zeitschrift für Geomorphologie* suppl. 36: 16–26.

Brunsden, D. 1990. Tablets of stone: toward the ten commandments of geomorphology. *Zeitschrift für Geomorphologie* suppl. 79: 1–37.

Brunsden, D. and Thornes, J. B. 1979. Landscape sensitivity and change. *Transactions of the Institute of British Geographers* n. s. 4: 463–84.

Brush, L. M. 1961. Drainage basins, channels, and flow characteristics of selected streams in central Pennsylvania. *US Geological Survey Professional Paper* 282-F.

Bull, W. B. 1991. *Geomorphic Responses to Climate Change*. New York: Oxford University Press.

Burrough, P. A. 1983. Multiscale sources of spatial variation in soils. *Journal of Soil Science* 34: 577–620.

Campbell, J. B. 1979. Spatial variability of soils. *Annals of the Association of American Geographers* 69: 544–56.

Carpenter, S. R. and Chaney, J. E. 1983. Scale of spatial patterns: four methods compared. *Vegetatio* 53: 153–60.

Carson, M. A. and Kirkby, M. J. 1972. *Hillslope Form and Process*. London: Cambridge University Press.

Carson, M. A. and LaPointe, M. F. 1983. The inherent asymmetry of river meander planform. *Journal of Geology* 91: 41–55.

Carter, R. W. G. and Orford, J. 1991. The sedimentary organisation and behavior of drift-aligned barriers. In *Coastal Sediments '91*. New York: American Society of Civil Engineers, pp. 934–48.

Catt, J. A. 1991. Soils as indicators of Quaternary climatic change in mid-latitude regions. *Geoderma* 51: 167–87.

Chang, P., Wang, B., Li, T., and Ji, L. 1994. Interactions between the seasonal cycle and the southern oscillation: frequency entrainment and chaos in a coupled ocean–atmosphere model. *Geophysical Research Letters* 21: 2817–20.

Chappell, J. 1983. Thresholds and lags in geomorphologic changes. *Australian Geographer* 15: 358–66.

Chase, C. G. 1992. Fluvial landsculpting and the fractal dimension of topography. *Geomorphology* 5: 39–57.

Chesworth, W. 1973. The parent rock effect in the genesis of soils. *Geoderma* 10: 215–25.

Chin, A. 1989. Step pools in stream channels. *Progress in Physical Geography* 13: 391–407.

Chorley, R. J. and Kennedy, B. A. 1971. *Physical Geography: a Systems Approach*. Englewood Cliffs, NJ: Prentice-Hall.

Christofolletti, A. 1993. Implicações geográficas relacionadas com as mudanças climáticas globais. *Boletin de Geográficas Teoretica* 23: 18–31.

Claps, P. and Oliveto, G. 1996. Reexamining the determination of the fractal dimension of river networks. *Water Resources Research* 32: 3123–35.

Clifford, N. J. 1993. Formation of riffle-pool sequences: field evidence for an autogenic process. *Sedimentary Geology* 85: 39–51.

Coles, S. M. 1979. Benthic microalgal populations on intertidal sediments and their role as precursors to salt marsh development. In R. L. Jeffreys, and A. J. Davy (eds), *Ecological Processes in Coastal Environments*. Oxford: Blackwell, pp. 25–42.

Committee on Opportunities in the Hydrologic Sciences. 1991. *Opportunities in the Hydrologic Sciences*. Washington: National Research Council, National Academy Press.

Crickmay, C. H. 1976. The hypothesis of unequal activity. In W. H. Melhorn and R. C. Flemal (eds), *Theories of Landform Development*. Boston: Allen and Unwin, pp. 103–9.

Culf, A. D., Allen, S. J., Gash, J. H. C., Lloyd, C. R., and Wallace, J. S. 1993. Energy and water budgets of an area of patterned woodland in the Sahel. *Agricultural and Forest Meteorology* 66: 65–80.

Culling, W. E. H. 1957. Multicyclic streams and the equilibrium theory of grade. *Journal of Geology* 65: 259–74.

Culling, W. E. H. 1986. Highly erratic spatial variability of soil pH on Iping Common, West Sussex. *Catena* 13: 81–98.

Culling, W. E. H. 1987. Equifinality: modern approaches to dynamical systems and their potential for geographical thought. *Transactions of the Institute of British Geographers* 12: 57–72.

Culling, W. E. H. 1988a. A new view of the landscape. *Transactions of the Institute of British Geographers* 13: 345–60.

Culling, W. E. H. 1988b. Dimension and entropy in the soil-covered landscape. *Earth Surface Processes and Landforms* 13: 619–48.

Culling, W. E. H. and Datko, M. 1987. The fractal geometry of the soil-covered landscape. *Earth Surface Processes and Landforms* 12: 369–85.

Daniels, R. B. and Gamble, E. E. 1967. The edge effect in some Ultisols in the North Carolina coastal plain. *Geoderma* 1: 117–24.

Daniels, R. B., Gamble, E. E., and Wheeler, W. H. 1978. Age of soil landscapes in the coastal plain of North Carolina. *Soil Science Society of America Journal* 42: 98–105.

Davies, T. R. H. and Sutherland, A. J. 1983. Extremal hypotheses for river behavior. *Water Resources Research* 19: 141–8.

Davis, W. M. 1889. Topographical development of the Triassic formation of the Connecticut Valley. *American Journal of Science* 37: 423–34.

Davis, W. M. 1899. The geographical cycle. *Geographical Journal* 14: 478–504.

Davis, W. M. 1902. Base-level, grade, and peneplain. *Journal of Geology* 10: 77–111.

Davis, W. M. 1909. *Geographical Essays*. Boston: Ginn.

DeBoer, D. H. 1992. Hierarchies and spatial scale in process geomorphology: a review. *Geomorphology* 4: 303–18.

Dekker, L. W. and Ritsema, C. J. 1994. Fingered flow: the creator of sand columns in dune and beach sands. *Earth Surface Processes and Land-forms* 19: 153–64.

Delcourt, H. R., Delcourt, P. A., and Webb, T. III 1983. Dynamic plant ecology: the spectrum of vegetational change in space and time. *Quaternary Science Reviews* 1: 153–75.

Demaree, G. R. and Nicolis, C. 1990. Onset of Sahelian drought viewed as a fluctuation-induced transition. *Quarterly Journal of the Royal Meteorological Society* 116: 221–38.

Dietrich, W. E., Wilson, C. J., Montgomery, D. R., McKean, J., and Bauer, R. 1992. Erosion thresholds and land surface morphology. *Geology* 20: 675–9.

Dunne, T. and Aubry, B. F. 1986. Evaluation of Horton's theory of sheetwash and rill erosion on the basis of field experiments. In A. D. Abrahams (ed.), *Hillslope Geomorphology*. Proceedings of the 17th Binghamton Geomorphology Symposium. London: Edward Arnold, pp. 31–53.

Dunne, T., Zhang, W., and Aubry, B. F. 1991. Effects of rainfall, vegetation, and microtopography on infiltration and runoff. *Water Resources Research* 27: 2271–85.

Eckmann, J-P. and Ruelle, D. 1985. Ergodic theory of chaos and strange attractors. *Reviews of Modern Physics* 54: 617–56.

Edmonds, W. J., Campbell, J. B., and Lentner, M. 1985. Taxonomic variation within three soil mapping units in Virginia. *Soil Science Society of America Journal* 49: 394–401.

Ehleringer, J. R. and Field, C. B. (eds). 1993. *Scaling Physiological Processes: Leaf to Globe*. San Diego: Academic Press.

Elliott, J. K. 1989. An investigation of the change in surface roughness through time on the foreland of Austre Okstindbreen, North Norway. *Computers and Geosciences* 15: 209–17.

Elsner, J. B. and Tsonis, A. A. 1993. Nonlinear dynamics established in the ENSO. *Geophysical Research Letters* 20: 213–16.

Engelen, G. G. and Venneker, R. G. W. 1988. ETA (erosion, transport, accumulation) systems, their classification, mapping, and management. In *Sediment Budgets*. Wallingford, UK: International Association of Hydrological Sciences Publication 174: 397–412.

Ergenzinger, P. 1987. Chaos and order: the channel geometry of gravel bed braided rivers. *Catena* suppl. 10: 85–98.

Ewing, L. K. and Mitchell, J. K. 1986. Overland flow and sediment transport simulation on small plots. *Transactions of the ASAE* 29: 1572–81.

Fasken, G. B. 1963. *Guide for Selecting Roughness Coefficient "n" Values for Channels*. Lincoln, NE: US Soil Conservation Service.

Ferguson, R. I. 1986. Hydraulics and hydraulic geometry. *Progress in Physical Geography* 10: 1–31.

Finke, P. A., Wösten, J. H. M., and Jansen, M. J. W. 1996. Effects of uncertainty in major input variables on simulated functional soil behavior. *Hydrological Processes* 10: 661–70.

Fiorentino, M. and Claps, P. 1992. On what can be explained by the entropy of a channel network. In V. P. Singh and M. Fiorentino (eds), *Entropy and Energy Dissipation in Water Resources*. Dordrecht: Kluwer, pp. 139–54.

Fiorentino, M., Claps, P. and Singh, V. P. 1992. An entropy-based morphological analysis of river basin networks. *Water Resources Research* 29: 1215–24.

Fraedrich, K. 1986. Estimating the dimensions of weather and climate attractors. *Journal of Atmospheric Sciences* 43: 410–32.

Fraedrich, K. 1987. Estimating weather and climate predictability on attractors. *Journal of Atmospheric Sciences* 44: 722–8.

Freidel, M. 1991. Range condition assessment and the concept of thresholds: a viewpoint. *Journal of Range Management* 44: 422–6.

Furley, P. A. 1996. The influence of slope on the nature and distribution of soils and plant communities in the central Brazilian cerrado. In M. G. Anderson and S. M. Brooks (eds), *Advances in Hillslope Processes*. Chichester: John Wiley, pp. 328–46.

Gaffin, S. R. and Maasch, K. A. 1991. Anomalous cyclicity in climate and stratigraphy and modeling nonlinear oscillations. *Journal of Geophysical Research* 96B: 6701–11.

Gardner, T. W., Jorgensen, D. W., Shuman, C., and Lemieux, C. R. 1987. Geomorphic and tectonic process rates: effects of measured time interval: *Geology* 15: 259–61.

Gerrard, A. J. 1992. *Soil Geomorphology*. London: Chapman and Hall.

Gerrard, A. J. 1993. Soil geomorphology: present dilemmas and future challenges. *Geomorphology* 7: 61–84.

Gersper, P. L. and Holowaychuck, N. 1970. Effects of stemflow water on a Miami soil under a beech tree. I: Morphological and physical properties. *Soil Science Society of America Proceedings* 34: 779–86.

Goodchild, M. F. and Klinkenberg, B. 1993. Statistics of channel networks on fractional Brownian surfaces. In N. S. Lam and L. De Cola (eds), *Fractals in Geography*. Englewood Cliffs, NJ: Prentice Hall, pp. 122–41.

Griffiths, G. A. 1984. Extremal hypotheses for river regime: an illusion of progress. *Water Resources Research* 20: 113–18.

Gupta, V. K. and Waymire, E. 1989. Statistical self-similarity in river networks parameterized by elevation. *Water Resources Research* 25: 463–76.

Hack, J. T. 1960. Interpretation of erosional topography in humid temperate regions. *American Journal of Science* 285: 80–97.

Hackney, C. T., Brady, S., Stemmy, L., Boris, M., Dennis, C., Hancock, T., O'Bryon, M., Tilton, C., and Barbee, E. 1996. Does intertidal vegetation indicate specific soil and hydrologic conditions? *Wetlands* 16: 89–94.

Haigh, M. J. 1987. The holon: hierarchy theory and landscape research. *Catena* suppl. 10: 181–92.

Haigh, M. J. 1988. Dynamic systems approaches in landslide hazard research. *Zeitschrift für Geomorphologie* suppl. 67: 79–91.

Haigh, M. J. 1989. Evolution of an anthropogenic desert gully system. In *Erosion, Transport, and Deposition Processes*. Wallingford, UK: International Association of Hydrological Sciences Publication 189: 65–77.

Hallet, B. 1990. Self-organization in freezing soils: from microscopic ice lenses to patterned ground. *Canadian Journal of Physics* 68: 842–52.

Hardisty, J. 1987. The transport response function and relaxation time in geomorphic modelling. *Catena* suppl. 10: 171–9.

Hardy, J. P. and Albert, M. R. 1995. Snow-induced thermal variations around a single conifer tree. *Hydrological Processes* 9: 923–34.

Harris, D., Fry, G. J., and Miller, S. T. 1994. Microtopography and agriculture in semi-arid Botswana. 2: Moisture availability, fertility, and crop performance. *Agricultural Water Management* 26: 133–48.

Harrison, J. B. J., McFadden, L. D., and Weldon, R. J. III 1990. Spatial soil variability in the Cajon Pass chronosequence: implications for the use of soils as a geochronological tool. In L. D. McFadden and P. L. K.

Knuepfer (eds), *Soils and Landscape Evolution*. Proceedings of the 21st Binghamton Symposium in Geomorphology. Amsterdam: Elsevier, pp. 391–8.

Helmlinger, K. R., Kumar, P., and Foufoula-Georgiou, E. 1993. On the use of digital elevation data for Hortonian and fractal analyses of channel networks. *Water Resources Research* 29: 2599–613.

Hendry, R. J. and McGlade, J. M. 1995. The role of memory in ecological systems. *Proceedings of the Royal Society of London* 259: 153–9.

Hey, R. D. 1978. Determinate hydraulic geometry of river channels. *Journal of the Hydraulics Division of the American Society of Civil Engineers* 104: 869–85.

Hey, R. D. 1979. Dynamic process–response model of river channel development. *Earth Surface Processes and Landforms* 4: 59–72.

Hill, D. E. and Parlange, J-Y. 1972. Wetting front instability in layered soils. *Soil Science Society of America Proceedings* 36: 697–702.

Hillel, D. and Baker, R. S. 1988. A descriptive theory of fingering during infiltration into layered soils. *Soil Science* 146: 51–6.

Hjelmfelt, A. T. 1988. Fractals and the river-length catchment-area ratio. *Water Resources Bulletin* 24: 455–9.

Hobbs, R. J. 1994. Dynamics of vegetation mosaics: can we predict responses to global change. *Ecoscience* 1: 346–56.

Holliday, V. T. 1994. The "state factor" approach in geoarcheology. In *Factors of Soil Formation: a Fiftieth Anniversary Retrospective*. Madison, WI: Soil Science Society of America Special Publication 33: 65–86.

Holtmeier, F-K. and Broll, G. 1992. The influence of tree islands and microtopography on pedoecological conditions in the forest-alpine tundra ecotone on Niwot Ridge, Colorado Front Range, USA. *Arctic and Alpine Research* 24: 216–28.

Hooke, J. M. and Redmond, C. E. 1992. Causes and nature of river planform change. In P. Billi, R. D. Hey, C. R. Thorne, and P. Tacconi (eds), *Dynamics of Gravel-bed Rivers*. Chichester: John Wiley, pp. 559–71.

Howard, A. D. 1971. Simulation of stream networks by headward growth and branching. *Geographical Analysis* 3: 29–50.

Howard, A. D. 1990. Theoretical model of optimal drainage networks. *Water Resources Research* 26: 2107–17.

Howard, A. D. 1994. A detachment-limited model of drainage basin evolution. *Water Resources Research* 30: 2261–85.

Hudson, B. D. 1992. The soil survey as pardigm-based science. *Soil Science Society of America Journal* 56: 836–41.

Huggett, R. J. 1975. Soil landscape system: a model of soil genesis. *Geoderma* 13: 1–22.

Huggett, R. J. 1985. *Earth Surface Systems*. Berlin: Springer.

Huggett, R. J. 1990. *Catastrophism: Systems of Earth History*. London: Edward Arnold.

Huggett, R. J. 1991. *Climate, Earth Processes, and Earth History*. Berlin: Springer.

Huggett, R. J. 1995. *Geoecology: an Evolutionary Approach*. London: Routledge.

Hupp, C. R. 1988. Plant ecological aspects of flood geomorphology. In V. R. Baker, R. C. Kochel, and P. C. Patton (eds), *Flood Geomorphology*. New York: Wiley, pp. 335–56.

Ibanez, J. J. 1994. Evolution of fluvial dissection landscapes in Mediterranean environments: quantitative estimates and geomorphic, pedologic, and phytocenotic repercussions. *Zeitschrift für Geomorphologie* 38: 105–19.

Ibanez, J. J., Ballexta, R. J., and Alvarez, A. G. 1990. Soil landscapes and drainage basins in Mediterranean mountain areas. *Catena* 17: 573–83.

Ibanez, J. J., De-Alba, S., Bermudez, F-F., and Garcia-Alvarez, A. 1995. Pedodiversity: concepts and measures. *Catena* 24: 215–32.

Ijjasz-Vasquez, E. J., Rodriguez-Iturbe, I., and Bras, R. L. 1992. On the multifractal characteristics of river basins. In J. D. Phillips and W. H. Renwick (eds), *Geomorphic Systems*. Amsterdam: Elsevier, pp. 297–310.

Jaffe, B. E. and Rubin, D. M. (1996) Using nonlinear forecasting to learn the magnitude and phasing of time-varying sediment suspension in the surf zone. *Journal of Geophysical Research* 101C: 14,283–96.

Jayawardena, A. W. and Lai, F. 1994. Analysis and prediction of chaos in rainfall and stream flow time series. *Journal of Hydrology* 153: 23–52.

Jenny, H. 1941. *Factors of Soil Formation: a System of Quantitative Pedology*. New York: McGraw-Hill.

Jenny, H. 1961. Derivation of state factor equations of soils and ecosystems. *Soil Science Society of America Proceedings* 25: 385–8.

Jenny, H. 1980. *The Soil Resource: Origin and Behavior*. New York: Springer.

Joffe, J. S. 1949. *Pedology*. New Brunswick, NJ: Pedology Publications.

Johnson, D. L. 1990. Biomantle evolution and the redistribution of earth materials and artifacts. *Soil Science* 149: 84–102.

Johnson, D. L. 1993. Dynamic denudation evolution of tropical, subtropical, and temperate landscapes with three-tiered soils: toward a general theory of landscape evolution. *Quaternary International* 17: 67–78.

Johnson, D. L. and Hole, F. D. 1994. Soil formation theory: a summary of its principal impacts on geography, geomorphology, soil-geomorphology, Quaternary geology, and paleopedology. In *Factors of Soil Formation: a Fiftieth Anniversity Retrospective*. Madison, WI: Soil Science Society of America Special Publication 33: 111–26.

Johnson, D. L. and Watson-Stegner, D. 1987. Evolution model of pedogenesis. *Soil Science* 143: 349–66.

Kapitaniak, T. 1988. *Chaos in Systems with Noise*. Singapore: World Scientific.

Karlinger, M. R. and Troutman, B. M. 1992. Fat fractal scaling of drainage networks from a random spatial network model. *Water Resources Research* 28: 1975–81.

Kearney, M. S. and Stevenson, J. C. 1991. Island land loss and marsh vertical accretion rate evidence for historical sea-level changes in Chesapeake Bay. *Journal of Coastal Research* 7: 403–15.

Kearney, M. S., Grace, R. E., and Stevenson, J. C. 1988. Marsh loss in the Nanticoke estuary, Chesapeake Bay. *Geographical Review* 78: 205–20.

Kearney, M. S., Stevenson, J. C., and Ward, L. G. 1994. Spatial and temporal changes in marsh vertical accretion rates: implications for sea-level rise. *Journal of Coastal Research* 10: 1010–20.

Kellert, S. H. 1993. *In the Wake of Chaos*. Chicago: University of Chicago Press.

Kempel-Eggenberger, C. 1993. Risse in der geoökologischen realität: chaos und ordnung in geoökologischen systemen. *Erdkunde* 47: 1–11.

Kiernan, K. 1990. Weathering as an indicator of the age of Quaternary glacial deposits in Tasmania. *Australian Geographer* 21: 1–17.

King, A. W. 1991. Translating models across scales in the landscape. In

M. G. Turner and R. H. Gardner (eds), *Quantitative Methods in Landscape Ecology*. New York: Springer, pp. 479–517.

King, L. C. 1953. Canons of landscape evolution. *Geological Society of America Bulletin* 64: 721–52.

Kirchner, J. W. 1993. Statistical inevitability of Horton's laws and the apparent randomness of stream channel networks. *Geology* 21: 591–4.

Kirchner, J. W. 1994. Reply (to comments on Kirchner 1993). *Geology* 22: 380–1.

Kirkby, M. J. 1971. Hillslope process–response models based on the continuity equation. *Institute of British Geographers Special Publication* 3: 15–30.

Kirkby, M. J. 1985. A model for the evolution of regolith-mantled slopes. In M. J. Woldenberg (ed.), *Models in Geomorphology*. London: Allen and Unwin, pp. 213–37.

Knighton, A. D. 1984. *Fluvial Forms and Processes*. London: Edward Arnold.

Knox, J. C. 1985. Responses of floods to Holocene climate change in the upper Mississippi Valley. *Quaternary Research* 23: 287–300.

Knox, J. C. 1993. Large increases in flood magnitude in response to modest changes in climate. *Nature* 361: 430–2.

Kraft, J. C. 1971. Sedimentary facies patterns and geologic history of an estuarine transgression. *Geological Society of America Bulletin* 82: 2131–58.

Kramer, S. and Marder, M. 1992. Evolution of river networks. *Physical Review Letters* 68: 205–8.

Kupfer, J. A. and Cairns, D. M. 1996. The suitability of montane ecotones as indicators of global climatic change. *Progress in Physical Geography* 20: 253–72.

La Barbera, P. and Rosso, R. 1989. On the fractal dimension of river networks. *Water Resources Research* 25: 735–41.

La Barbera, P. and Roth, G. 1994. Invariance and scaling parameters in the distributions of contributing area and energy in drainage basins. *Hydrological Processes* 8: 125–35.

Lamberti, A. 1988. About extremal hypotheses and river regime. In W. B. White, (ed.), *International Conference on River Regime*. New York: John Wiley, pp. 121–34.

Lamberti, A. 1992. Dynamic and variational approaches to the river regime relation. In V. P. Singh and M. Fiorentino (eds), *Entropy and Energy Dissipation in Water Resources*. Dordrecht: Kluwer, pp. 507–25.

Langbein, W. B. 1964. Geometry of river channels. *Journal of the Hydraulics Division of the American Society of Civil Engineers* 90: 301–12.

Lapenis, A. G. and Shabalova, M. V. 1994. Global climate changes and moisture conditions in the intercontinental arid zones. *Climatic Change* 27: 283–97.

Lare, A. R. and Nicholson, S. E. 1994. Contrasting conditions of surface water balance in wet years and dry years as a possible land surface–atmosphere feedback mechanism in the west African Sahel. *Journal of Climate* 7: 653–701.

Laycock, W. 1991. Stable states and thresholds of range condition on North American rangelands: a viewpoint. *Journal of Range Management* 44: 427–31.

Leopold, L. B. and Langbein, W. B. 1962. The concept of entropy in landscape evolution. *US Geological Survey Professional Paper* 500A.

Leopold, L. B. and Maddock, T. 1953. The hydraulic geometry of stream channels and some physiographic implications. *US Geological Survey Professional Paper* 252: 1–57.

Levin, S. A. 1992. The problem of pattern and scale in ecology. *Ecology* 73: 1943–67.

Li, W. 1990. Mutual information functions versus correlation functions. *Journal of Statistical Physics* 60: 823–37.

Liu, H-S. 1995. A new view on the driving mechanism of Milankovitch glaciation cycles. *Earth and Planetery Science Letters* 131: 17–26.

Liu, T. 1992. Fractal structure and properties of stream networks. *Water Resources Research* 28: 2981–8.

Liu, Y., Steenhuis, T. S., and Parlange, J-Y. 1994. Formation and persistence of fingered flow fields in coarse grained soils under different moisture contents. *Journal of Hydrology* 159: 187–95.

Lockwood, J. G. 1986. The causes of drought with particular reference to the Sahel. *Progress in Physical Geography* 10: 111–19.

Loewenherz, D. S. 1991. Stability and the initiation of channelized surface drainage: a reassessment of the short wavelength limit. *Journal of Geophysical Research* 96B: 8453–564.

Logofet, D. O. 1993. *Matrices and Graphs: Stability Problems in Mathematical Ecology*. Boca Raton, FL: CRC Press.

Lorenz, E. N. 1963. Deterministic nonperiodic flow. *Journal of Atmospheric Sciences* 20: 130–41.

Lorenz, E. N. 1965. A study of the predictability of a 28-variable atmospheric model. *Tellus* 27: 321–33.

Lorenz, E. N. 1990. Can chaos and intransitivity lead to interannual variability? *Tellus* 42A: 378–89.

McAuliffe, J. R. 1994. Landscape evolution, soil formation, and ecological patterns and processes in Sonoran desert bajadas. *Ecological Monographs* 64: 111–48.

McBratney, A. B. 1992. On variation, uncertainty, and informatics in environmental soil management. *Australian Journal of Soil Research* 30: 913–36.

McDonald, E. V. and Busacca, A. J. 1990. Interaction between aggrading geomorphic surfaces and the formation of a Late Pleistocene paleosol in the Palouse loess of eastern Washington state. In L. D. McFadden and P. L. K. Knuepfer (eds), *Soils and Landscape Evolution*. Proceedings of the 21st Binghamton Symposium in Geomorphology. Amsterdam: Elsevier, pp. 449–70.

McKee, E. D. (ed.) 1979. *A Study of Global Sand Seas*. Washington: US Geological Survey Professional Paper 1052.

McSweeney, K., Gessler, P. E., Slater, B., Hammer, R. D., Bell, J., and Petersen, G. W. 1994. Towards a new framework for modeling the soil-landscape continuum. In *Factors of Soil Formation: a Fiftieth Anniversity Retrospective*. Madison, WI: Soil Science Society of America Special Publication 33: 127–46.

Mainguet, M. 1991. *Desertification: Natural Background and Human Mismanagement*. Berlin: Springer.

Major, J. 1951. A functional factorial approach to plant ecology. *Ecology* 32: 392–412.

Malanson, G. P., Butler, D. R., and Georgakakos, K. P. 1992. Nonequilibrium geomorphic processes and deterministic chaos. In Phillips and Renwick, 1992, pp. 311–22.

Marani, A., Rigon, R., and Rinaldo, A. 1991. A note on fractal channel networks. *Water Resources Research* 27: 3041–9.

Maritan, A., Colatori, F., Flammini, A., Cieplak, M., and Banavar, J. R. 1996. Universality classes of optimal channel networks. *Science* 272: 984–6.

Markewich, H. W., Pavich, M. J., Mausbach, M. J., Johnson, R. G., and Gonzalez, V. M. 1989. *A Guide for Using Soil and Weathering Profile Data in Chronosequence Studies of the Coastal Plain of the Eastern United States.* Washington: US Geological Survey Bulletin 1589-D.

Martin, P. 1993. Vegetation responses and feedbacks to climate: a review of models and processes. *Climate Dynamics* 8: 201–10.

Martini, I. P. and Chesworth, W. (eds). 1992. *Weathering, Soils, and Paleosols.* Amsterdam: Elsevier.

Masek, J. G. and Turcotte, D. L. 1993. A diffusion-limited aggregation model for the evolution of drainage networks. *Earth and Planetary Science Letters* 119: 379–86.

Mausbach, M. J., Brasher, B. R., Yeck, R. D., and Nettleton, W. D. 1980. Variability of measured properties in morphologically matched pedons. *Soil Science Society of America Journal* 44: 358–63.

May, R. M. 1973. *Stability and Complexity in Model Ecosystems.* Princeton, NJ: Princeton University Press.

May, R. M. 1976. Simple mathematical models with very complicated dynamics. *Nature* 261: 459–67.

Meakin, P. 1991. Fractal aggregates in geophysics. *Reviews of Geophysics* 29: 317–54.

Middleton, N. and Thomas, D. S. G. 1992. *World Atlas of Desertification.* United Nations Environment Program. London: Edward Arnold.

Miller, D. J. and Dunne, T. 1996. Topographic perturbations of regional stresses and consequent bedrock fracturing. *Journal of Geophysical Research* 101B: 25,523–36.

Miller, S. T., Brinn, P. J., Fry, G. J., and Harris, D. 1994. Microtopography and agriculture in semi-arid Botswana. 1: Soil variability. *Agricultural Water Management* 26: 107–32.

Miller, T. K. 1991. An assessment of the equable change principle in at-a-station hydraulic geometry. *Water Resources Research* 27: 2751–8.

Moglen, G. E. and Bras, R. L. 1995. The effect of spatial heterogeneities on geomorphic expression in a model of basin evolution. *Water Resources Research* 31: 2613–23.

Mohanty, B. P., Ankeny, M. D., Horton, R., and Kanwar, R. S. 1994. Spatial analysis of hydraulic conductivity measured using disc infiltrometer. *Water Resources Research* 30: 2489–98.

Mossa, J. and Schumacher, B. A. 1993. Fossil tree casts in south Louisiana soils. *Journal of Sedimentary Petrology* 63: 707–13.

Muhs, D. R. 1984. Intrinsic thresholds in soil systems. *Physical Geography* 5: 99–110.

Nahon, D. B. 1991. Self-organization in chemical lateritic weathering. *Geoderma* 51: 5–13.

Nanson, G. C. and Erskine, W. D. 1988. Episodic changes of channels and floodplains on coastal rivers in New South Wales. In R. F. Warner (ed.), *Fluvial Geomorphology of Australia.* Sydney: Academic Press, pp. 201–22.

Nelson, J. M. 1990. The initial instability and finite-amplitude stability of alternate bars in straight channels. *Earth Science Reviews* 29: 97–115.

Nicholl, M. J., Wheatcraft, S. W., Tyler, S. W., and Berkowitz, B. 1994. Is Old Faithful a strange attractor? *Journal of Geophysical Research* 99B: 4495–503.

Nichols, M. M. 1989. Sediment accumulation rates and relative sea-level rise in lagoons. *Marine Geology* 88: 201–19.

Nicolau, J. M., Solé-Benet, A., Puigdefábregas, J., and Gutiérrez, L. 1996. Effects of soil and vegetation on runoff along a catena in semi-arid Spain. *Geomorphology* 14: 297–309.

Nicolis, G. and Nicolis, G. 1984. Is there a climatic attractor? *Nature* 311: 529–32.

Nijkamp, P. and Reggiani, A. 1992. Impacts of multiple-period lags in dynamic logit models. *Geographical Analysis* 24: 159–73.

Nikiforoff, C. C. 1959. Reappraisal of the soil. *Science* 129: 186–96.

Nikora, V. I. 1991. Fractal structures of river plan forms. *Water Resources Research* 27: 1327–33.

Nikora, V. I. 1994. On self-similarity and self-affinity of drainage basins. *Water Resources Research* 39: 133–7.

Nikora, V. I., Ibbitt, R., and Shankar, U. 1996. On channel network fractal properties: a case study of the Hutt River basin, New Zealand. *Water Resources Research* 32: 3375–84.

Nortcliff, S. 1978. Soil variability and reconnaissance soil mapping: a statistical study in Norfolk. *Journal of Soil Science* 29: 403–18.

Nyman, J. A., DeLaune, R. D., Roberts, H. H., and Patrick, W. H. 1993. Relationship between vegetation and soil formation in a rapidly submerging coastal marsh. *Marine Ecology Progress Series* 96: 264–79.

Nyman, J. A., Carloss, M., DeLaune, R. D., and Patrick, W. H. 1994. Erosion rather than plant dieback as the mechanism of marsh loss in an estuarine marsh. *Earth Surface Processes and Landforms* 19: 69–84.

Oliver, M. A. and Webster, R. 1986a. Semivariograms for modelling the spatial pattern of landform and soil properties. *Earth Surface Processes and Landforms* 11: 491–504.

Oliver, M. A. and Webster, R., 1986b. Combining nested and linear sampling for determining the scale and form of spatial variation of regionalized variables. *Geographical Analysis* 18: 227–42.

Ollier, C. and Pain, C. F. 1996. *Regolith, Soils, and Landforms*. Chichester: John Wiley.

O'Neill, R. V. 1988. Hierarchy theory and global change. In T. Rosswall, R. G. Woodmansee, and R. G. Risser (eds), Scales and Global Change. Chichester: Wiley, pp. 29–45.

O'Neill, R. V. 1989. Perspectives in hierarchy and scale. In J. Roughgarden, R. M. May, and S. A. Levin (eds), *Perspectives in Ecological Theory*. Princeton, NJ: Princeton, University Press, pp. 140–56.

O'Neill, R. V., DeAngelis, D. L., Waide, J. B., and Allen, T. F. H. 1986. *A Hierarchical Concept of Ecosystems*. Princeton, NJ: Princeton University Press.

Oono, Y. 1978. A heuristic approach to the Kolmogorov entropy as a disorder parameter. *Progress in Theoretical Physics* 60: 1944–6.

Orford, J. D., Carter, R. W. G., and Jennings, S. C. 1991. Coarse clastic barrier environments: evolution and implications for Quaternary sea level interpretation. *Quaternary International* 9: 87–104.

Orson, R. A. and Howes, B. L. 1992. Salt marsh development studies at Waquoit Bay, Massachusetts: influence of geomorphology on long-term plant community structure. *Estuarine, Coastal, and Shelf Science* 35: 453–71.

Orson, R. A., Panagetou, W., and Leatherman, S. P. 1985. Response of tidal salt marshes of the US Atlantic and Gulf coasts to rising sea levels. *Journal of Coastal Research* 1: 29–37.

Orson, R. A., Simpson, R. L., and Good, R. E. 1992. The paleoecological development of a late Holocene, tidal freshwater marsh of the upper Delaware River estuary. *Estuaries* 15: 130–46.

Osterkamp, W. R. and Hupp, C. R. 1996. The evolution of geomorphology, ecology, and other composite sciences. In B. Rhoads and C. Thorn (eds), *The Scientific Nature of Geomorphology*. New York: John Wiley, pp. 415–41.

Otto-Bleisner, B. L. 1995. Continental drift, runoff, and weathering feedbacks: implications from climate model experiments. *Journal of Geophysical Research* 100D: 11,537–48.

Pahl-Wostl, C. 1995. *The Dynamic Nature of Ecosystems: Chaos and Order Entwined*. Chichester: John Wiley.

Palmer, T. N. 1993. Extended range atmospheric prediction and the Lorenz model. *Bulletin of the American Meteorological Society* 74: 49–65.

Parker, K. C. 1995. Effects of complex geomorphic history on soil and vegetation patterns on arid alluvial fans. *Journal of Arid Environments* 30: 19–39.

Parker, K. C. and Bendix, J. 1996. Landscape-scale geomorphic influences on vegetation patterns in four environments. *Physical Geography* 17: 113–41.

Parsons, A. J., Abrahams, A. D., and Wainwright, J. 1996. Responses of interrill runoff and erosion rates to vegetation change in southern Arizona. *Geomorphology* 14: 311–17.

Parton, W. J., Schimel, D. S., Cole, C. V., and Ojima, D. S. 1987. Analysis of factors controlling soil organic matter levels in Great Plains grasslands. *Soil Science Society of America Journal* 51: 1173–9.

Paton, T. R. 1978. *The Formation of Soil Material*. London: Allen and Unwin.

Paton, T. R., Humphreys, G. S., and Mitchell, P. B. 1995. *Soils: a New Global View*. New Haven, CT: Yale University Press.

Penck, W. 1924. *Die Morphologische Analyse*. English translation by H. Czech, and K. C. Boswell 1953. New York: St Martin's Press.

Pennington, W. 1986. Lags in adjustment of vegetation to climate caused by the pace of soil development. *Vegetatio* 67: 105–18.

Perkins, J. S. and Thomas, D. S. G. 1993. Environmental responses and sensitivity to permanent cattle ranching, semi-arid western central Botswana. In D. S. G. Thomas and R. J. Allison (eds), *Landscape Sensitivity*. Chichester: John Wiley, pp. 273–86.

Perring, F. 1958. A theoretical approach to a study of chalk grassland. *Journal of Ecology* 46: 665–79.

Pethick, J. 1974. The distribution of salt pans on tidal salt marshes. *Journal of Biogeography* 1: 57–62.

Phillips, J. D. 1985. Stability of artificially drained lowlands: a theoretical assessment. *Ecological Modelling* 27: 69–79.

Phillips, J. D. 1986a. Coastal submergence and marsh fringe erosion. *Journal of Coastal Research* 2: 427–36.

Phillips, J. D. 1986b. Sediment storage, sediment yield, and time scales in landscape denudation studies. *Geographical Analysis* 18: 161–7.

Phillips, J. D. 1986c. Spatial analysis of shoreline erosion, Delaware Bay, New Jersey. *Annals of the Association of American Geographers* 76: 50–62.

Phillips, J. D. 1986d. Measuring complexity of environmental gradients. *Vegetatio* 64: 95–102.

Phillips, J. D. 1987a. Sediment budget stability in the Tar River basin, North Carolina. *American Journal of Science* 287: 780–94.

Phillips, J. D. 1987b. Choosing the level of detail for depicting two-variable spatial relationships. *Mathematical Geology* 19(6): 539–47.

Phillips, J. D. 1988a. Nonpoint source pollution and spatial aspects of risk assessment. *Annals of the Association of American Geographers* 78: 611–23.

Phillips, J. D. 1988b. The role of spatial scale in geomorphic systems. *Geographical Analysis* 20: 359–68.

Phillips, J. D. 1988c. Incorporating fluvial change in hydrologic simulations: A case study in coastal North Carolina. *Applied Geography* 8: 25–36.

Phillips, J. D. 1989a. Fluvial sediment storage in wetlands. *Water Resources Bulletin* 25: 867–73.

Phillips, J. D. 1989b. Estimating minimum achievable soil loss in developing countries. *Applied Geography* 9: 219–36.

Phillips, J. D. 1989c. Nonpoint source pollution control effectiveness of riparian forests along a coastal plain river. *Journal of Hydrology* 110: 221–37.

Phillips, J. D. 1989d. An evaluation of the state factor model of soil ecosystems. *Ecological Modelling* 45: 165–77.

Phillips, J. D. 1989e. Hillslope and channel sediment delivery and impacts of soil erosion on water resources. In R. F. Hadley (ed.), *Sediment and the Environment*. Wallingford, UK: International Association of Hydrological Sciences Publication 184: 183–90.

Phillips, J. D. 1989f. An evaluation of the factors determining the effectiveness of water quality buffer zones. *Journal of Hydrology* 107: 133–45.

Phillips, J. D. 1989g. Erosion and planform irregularity of an estuarine shoreline. *Zeitschrift für Geomorphologie suppl.* 73: 59–71.

Phillips, J. D. 1990a. Relative ages of wetland and upland surfaces as indicated by pedogenic development. *Physical Geography* 11: 363–78.

Phillips, J. D. 1990b. A saturation-based model for wetland identification. *Water Resources Bulletin* 26: 333–42.

Phillips, J. D. 1990c. Relative importance of factors influencing fluvial soil loss at the global scale. *American Journal of Science* 290: 547–68.

Phillips, J. D. 1990d. The instability of hydraulic geometry. *Water Resources Research* 26: 739–44.

Phillips, J. D. 1991a. The human role in earth surface systems: some theoretical considerations. *Geographical Analysis* 23: 316–31.

Phillips, J. D. 1991b. Fluvial sediment budgets in the North Carolina Piedmont. *Geomorphology* 4: 231–41.

Phillips, J. D. 1991c. Fluvial sediment delivery to a coastal plain estuary in the Atlantic drainage of the United States. *Marine Geology* 98: 121–34.

Phillips, J. D. 1991d. Multiple modes of adjustment in unstable river channel cross-sections. *Journal of Hydrology* 123: 39–49.

Phillips, J. D. 1992a. Delivery of upper-basin sediment to the lower Neuse River, North Carolina, USA. *Earth Surface Processes and Landforms* 17: 699–709.

Phillips, J. D. 1992b. The end of equilibrium? In Phillips and Renwick, 1992, pp. 195–201.

Phillips, J. D. 1992c. Deterministic chaos in surface runoff. In A. J. Parsons and A. D. Abrahams (eds), *Overland Flow: Hydraulics and Erosion Mechanics*. London: UCL Press, pp. 177–97.

Phillips, J. D. 1992d. Nonlinear dynamical systems in geomorphology: revolution or evolution? In Phillips and Renwick, 1992, pp. 219–29.

Phillips, J. D. 1992e. Qualitative chaos in geomorphic systems, with an example from wetland response to sea level rise. *Journal of Geology* 100: 365–74.

Phillips, J. D. 1992f. The source of alluvium in large rivers of the lower coastal plain of North Carolina. *Catena* 19: 59–75.

Phillips, J. D. 1993a. Chaotic evolution of some coastal plain soils. *Physical Geography* 14: 566–80.

Phillips, J. D. 1993b. Biophysical feedbacks and the risks of desertification. *Annals of the Association of American Geographers* 83: 630–40.

Phillips, J. D. 1993c. Progessive and regressive pedogenesis and complex soil evolution. *Quaternary Research* 40: 169–76.

Phillips, J. D. 1993d. Stability implications of the state factor model of soils as a nonlinear dynamical system. *Geoderma* 58: 1–15.

Phillips, J. D. 1993e. Pre- and post-colonial sediment sources and storage in the lower Neuse River basin, North Carolina. *Physical Geography* 14: 272–84.

Phillips, J. D. 1993f. Spatial-domain chaos in landscapes. *Geographical Analysis* 25: 101–17.

Phillips, J. D. 1993g. Interpreting the fractal dimension of river networks. In N. S. Lam and L. De Cola (eds), *Fractals in Geography*. Englewood Cliffs, NJ: Prentice-Hall, pp. 142–57.

Phillips, J. D. 1993h. Instability and chaos in hillslope evolution. *American Journal of Science* 293: 25–48.

Phillips, J. D. 1994. Deterministic uncertainty in landscapes. *Earth Surface Processes and Landforms* 19: 389–401.

Phillips, J. D. 1995a. Self-organization and landscape evolution. *Progress in Physical Geography* 19: 309–21.

Phillips, J. D. 1995b. Nonlinear dynamics and the evolution of relief. *Geomorphology* 14: 57–64.

Phillips, J. D. 1995c. Biogeomorphology and landscape evolution: the problem of scale. *Geomorphology* 13: 337–47.

Phillips, J. D. 1995d. Decoupling of sediment sources in large river basins. In W. R. Osterkamp (ed.), *Effects of Scale on Interpretation and Management of Sediment and Water Quality*. Wallingford, UK: International Association of Hydrological Sciences Publication 226: 11–16.

Phillips, J. D. 1995e. Time lags and emergent stability in morphogenic/pedogenic system models. *Ecological Modelling* 78: 267–76.

Phillips, J. D. 1996a. Deterministic complexity, explanation, and predictability in geomorphic systems. In B. Rhoads and C. Thorn (eds), *The Scientific Nature of Geomorphology*. Proceedings of the 27th Binghamton Geomorphology Symposium. New York: John Wiley, pp. 315–36.

Phillips, J. D. 1996b. Wetland buffers and runoff hydrology. In B. G. Warner and G. Mulamoottil (eds), *Wetlands: Buffers, Boundaries, and Gradients*. Boca Raton, FL: Lewis Publishers, pp. 207–20.

Phillips, J. D. 1997a. Simplexity and the reinvention of equifinality. *Geographical Analysis* 29: 1–15.

Phillips, J. D. 1997b. Humans as geological agents and the question of scale. *American Journal of Science* 298: 98–115.

Phillips, J. D. 1997c. Human agency, Holocene sea level, and floodplain accretion in coastal plain rivers. *Journal of Coastal Research* 13: 854–66.

Phillips, J. D. and Holder, G. R. 1991. Large organic debris in the lower Tar River, North Carolina, 1879–1900. *Southeastern Geographer* 31: 55–66.

Phillips, J. D. and Renwick, W. H. (eds) 1992. *Geomorphic Systems.* Proceedings of the 23rd Binghamton Geomorphology Symposium. Amsterdam: Elsevier.

Phillips, J. D. and Renwick, W. H. 1996. Surface instability and human modification in geomorphic systems. In I. Douglas, R. Huggett and M. Robinson (eds), *Encyclopedia of Geography.* London: Routledge, pp. 553–72.

Phillips, J. D. and Steila, D. 1984. Hydrologic equilibrium status of a disturbed eastern North Carolina watershed. *GeoJournal* 9: 351–7.

Phillips, J. D., Wyrick, M. J., Robbins, J. G., and Flynn, F. 1993. Accelerated erosion on the North Carolina coastal plain. *Physical Geography* 14: 114–30.

Phillips, J. D., Gosweiler, J., Tollinger, M., Mayeux, S., Gordon, R., Altieri, T., and Wittmeyer, M. 1994. Edge effects and spatial variability in coastal plain Ultisols. *Southeastern Geographer* 34: 125–37.

Phillips, J. D., Perry, D. C., Garbee, A. R., Carey, K. L., Stein, D., Morde, M. B., and Sheehy, J. 1996. Deterministic uncertainty and complex pedogenesis in some Pleistocene dune soils. *Geoderma* 73: 147–64.

Phillips, J. D., Lampe, M., King, R. T., Cedillo, M., Beachley, R., and Grantham, C. 1997. Ferricrete formation in the North Carolina coastal plain. *Zeitschrift für Geomorphologie* 41: 67–81.

Phipps, M. 1981. Entropy and community pattern analysis. *Journal of Theoretical Biology* 93: 253–73.

Pickup, G. 1988. Modelling arid zone soil erosion at the regional scale. In R. F. Warner (ed.), *Fluvial Geomorphology of Australia.* Sydney: Academic Press, pp. 105–28.

Pickup, G. and Chewings, V. H. 1986. Random field modelling of spatial variations in erosion and deposition in flat alluvial landscapes in arid central Australia. *Ecological Modelling* 33: 269–96.

Pielke, R. A. and Zeng, X. 1994. Long-term variability of climate. *Journal of the Atmospheric Sciences* 51: 155–9.

Pope, G. A., Dorn, R. I., and Dixon, J. C. 1995. A new conceptual model for understanding geographical variations in weathering. *Annals of the Association of American Geographers* 85: 38–64.

Price, A. G. 1994. Measurement and variability of physical properties and soil water distribution in a forest podzol. *Journal of Hydrology* 161: 347–64.

Procaccia, I. 1988. Weather systems: complex or just complicated? *Nature* 333: 498–9.

Prosser, I. P. and Roseby, S. J. 1995. A chronosequence of rapid leaching of mixed podzol soil materials following sand mining. *Geoderma* 64: 297–308.

Puccia, C. J. and Levins, R. 1985. *Qualitative Modeling of Complex Systems.* Cambridge, MA: Harvard University Press.

Puccia, C. J. and Levins, R. 1991. Qualitative modeling in ecology: loop analysis, signed digraphs, and time averaging. In P. A. Fishwick and P. A. Luker (eds), *Qualitative Simulation Modeling and Analysis.* New York: Springer, pp. 119–43.

Puigdefábregas, J. and Sánchez, G. 1996. Geomorphological implications of vegetation patchiness on semi-arid slopes. In M. G. Anderson and

S. M. Brooks (eds), *Advances in Hillslope Processes*, vol. 2. Chichester: John Wiley, pp. 1027–60.

Rampino, M. R. and Sanders, J. E. 1981. Episodic growth of Holocene tidal marshes in the Northeastern US: a possible indicator of eustatic sea-level fluctuations. *Geology* 9: 63–7.

Reed, D. J. 1990. The impact of sea level rise on coastal salt marshes. *Progress in Physical Geography* 14: 465–81.

Retallack, G. J. 1990. *Soils of the Past: an Introduction to Paleopedology*. London: Unwin Hyman.

Retallack, G. J. 1994a. The environmental factor approach to the interpretation of paleosols. In *Factors of Soil Formation: a Fiftieth Anniversary Retrospective*. Madison, WI: Soil Science Society of America Special Publication 33: 31–64.

Retallack, G. J. 1994b. A pedotype approach to the latest Cretaceous and earliest Tertiary paleosols in eastern Montana. *Geological Society of America Bulletin* 106: 1377–97.

Rey, V., Dames, A. G., and Belzons, M. 1995. On the formation of bars by the action of waves on an erodible bed: a laboratory study. *Journal of Coastal Research* 11: 1180–94.

Rhoads, B. L. 1988. Mutual adjustments between process and form in a desert mountain fluvial system. *Annals of the Association of American Geographers* 78: 271–87.

Rhoads, B. L. and Welford, M. R. 1991. Initiation of river meandering. *Progress in Physical Geography* 15: 127–56.

Richards, K. S. 1982. *Rivers: Form and Process in Alluvial Channels*. London: Methuen.

Richter, D. D. and Markewitz, D. 1995. How deep is the soil? *Bioscience* 45: 600–9.

Ridenour, G. S. and Giardino, J. R. 1991. The statistical study of hydraulic geometry: a new direction for compositional data analysis. *Mathematical Geology* 23: 349–66.

Riggs, S. R., Cleary, W. J., and Snyder, S. W. 1995. Influence of inherited geologic framework on barrier shoreface morphology and dynamics. *Marine Geology* 126: 213–34.

Rigon, R., Rinaldo, A., and Rodriguez-Iturbe, I. 1994. On landscape self-organization. *Journal of Geophysical Research* 99B: 11,971–93.

Rinaldo, A., Rodriguez-Iturbe, I., Rigon, R., Bras, R. L., Ijjasz-Vasquez, E. J., and Marani, A. 1992. Minimum energy and fractal structures of drainage networks. *Water Resources Research* 28: 2183–95.

Rinaldo, A., Rodriguez-Iturbe, I., Bras, R. L., and Ijjasz-Vasquez, E. J. 1993. Self-organized fractal river networks. *Physical Review Letters* 70: 822–66.

Ritsema, C. J. and Dekker, L. W. 1994. Soil moisture and dry bulk density patterns in bare dune sands. *Journal of Hydrology* 154: 107–32.

Ritsema, C. J. and Dekker, L. W. 1995. Distribution flow: a general process in the top layer of water repellent soils. *Water Resources Research* 31: 1187–200.

Ritsema, C. J., Dekker, L. W., Hendricks, J. M. H., and Hamminga, W. 1993. Preferential flow mechanism in a water repellent sandy soil. *Water Resources Research* 29: 2183–93.

Robertson, A. 1994. Directionality, fractals, and chaos in wind-shaped forests. *Agricultural and Forest Meteorology* 72: 133–66.

Rodriguez-Iturbe, I., Entekhabi, D., Lee, J-S., and Bras, R. L. 1991. Nonlinear dynamics of soil moisture at climate scales. 2: Chaotic analysis. *Water Resources Research* 27: 1899–915.

Rodriguez-Iturbe, I., Rinaldo, A., Rigon, R., Bras, R. L., Ijjasz-Vasquez, E. J., and Marani, A. 1992. Fractal structures as least energy patterns: the case of river networks. *Geophysical Research Letters* 19: 889–92.

Rosenstein, M. T., Collins, J. J., and DeLuca, C. J. 1993. A practical method for calculating largest Lyapunov exponents from small data sets. *Physica D* 65: 117–34.

Rosenzweig, M. L. 1995. *Species Diversity in Space and Time*. New York: Cambridge University Press.

Rosso, R., Bacchi, B., and La Barbera, P. 1991. Fractal relation of main-stream length to catchment area in river networks. *Water Resources Research* 27: 381–7.

Rosswall, T., Woodmansee, R. G., and Risser, P. G. (eds) 1988. *Scales and Global Change: Spatial and Temporal Variability in Biospheric and Geospheric Processes*. Chichester: John Wiley.

Rubin, D. M. 1992. Use of forecasting signatures to help distinguish periodicity, randomness, and chaos in ripples and other spatial patterns. *Chaos* 2: 525–35.

Ruhe, R. V. 1952. Topographic discontinuities of the Des Moines lobe. *American Journal of Science* 250: 46–56.

Runge, C. A. 1973. Soil development sequences and energy models. *Soil Science* 115: 183–93.

Russo, D. and Bresler, E. 1981. Soil hydraulic properties as stochastic processes. I: An analysis of field spatial variability. *Soil Science Society of America Journal* 45: 682–7.

Ryzhova, I. M. 1996. Analysis of the feedback effects of ecosystems produced by changes in carbon-cycling parameters using mathematical models. *Eurasian Soil Science* 28: 44–52.

Sangoyomi, T. B., Lall, U., and Abarbanel, H. D. I. 1996. Nonlinear dynamics of the Great Salt Lake: dimension estimation. *Water Resources Research* 32: 149–59.

Sapozhnikov, V. B. and Foufoula-Georgiou, E. 1996. Do the current landscape evolution models show self-organized criticality? *Water Resources Research* 32: 1109–12.

Savenjie, H. H. G. 1995. New definitions for moisture recycling and the relationship with land-use changes in the Sahel. *Journal of Hydrology* 167: 57–78.

Scatena, F. N. and Lugo, A. E. 1995. Geomorphology, disturbance, and the soil and vegetation of two subtropical wet steepland watersheds of Puerto Rico. In C. R. Hupp, W. R. Osterkamp and A. D. Howard (eds), *Biogeomorphology, Terrestrial and Freshwater Systems*. Amsterdam: Elsevier, pp. 199–214.

Schaetzl, R. J. 1986a. Soilscape analysis of contrasting glacial terrains in Wisconsin. *Annals of the Association of American Geographers* 76: 414–25.

Schaetzl, R. J. 1986b. Complete soil profile inversion by tree uprooting. *Physical Geography* 7: 181–9.

Schaetzl, R. J. 1990. Effects of treethrow microtopography on the characteristics and genesis of Spodosols, Michigan, USA. *Catena* 17: 111–26.

Schaetzl, R. J. and Follmer, L. R. 1990. Longevity of treethrow microtopography: implications for mass wasting. *Geomorphology* 3: 113–23.

Schaetzl, R. J., Johnson, D. L., Burns, S. F., and Small, T. W. 1989. Tree

uprooting: review of terminology, process, and environmental implications. *Canadian Journal of Forest Research* 19: 1–11.

Schaetzl, R. J., Burns, S. F., Small, T. W., and Johnson, D. L. 1990. Tree
uprooting: review of types and patterns of soil disturbance. *Physical
Geography* 11: 277–91.

Schaffer, W. M. 1981. Ecological abstraction: the consequences of reduced
dimensionality in ecological models. *Ecological Monographs* 51: 383–
401.

Schaffer, W. M. 1985. Order and chaos in ecological systems. *Ecology* 66:
93–106.

Scheidegger, A. E. 1967. A stochastic model for drainage patterns into an
intermontane trench. *Hydrological Sciences Bulletin* 12: 15–20.

Scheidegger, A. E. 1983. The instability principle in geomorphic equilibrium. *Zeitschrift für Geomorphologie* 27: 1–19.

Scheidegger, A. E. 1987. The fundamental principles of landscape evolution. *Gatena* suppl. 10: 199–210.

Scheidegger, A. E. 1990. *Theoretical Geomorphology*. Berlin: Springer.

Scheiner, S. M. and Rey-Benayas, J. M. 1994. Global patterns of plant
diversity. *Evolutionary Ecology* 8: 331–47.

Schlesinger, W. H., Reynolds, J. F., Cunningham, G. L., Huenneke, L. F.,
Jarrell, W. M., Virginia, R. A., and Whitford, W. G. 1990. Biological
feedbacks in global desertification. *Science* 247: 1043–8.

Schlesinger, W. H., Raikes, J. A., Hartley, R. E., and Cross, A. F. 1996. On
the spatial pattern of soil nutrients in desert ecosystems. *Ecology* 77:
364–74.

Schumm, S. A. 1956. The evolution of drainage systems and slopes in
badlands at Perth Amboy, New Jersey. *Geological Society of America
Bulletin* 67: 597–646.

Schumm, S. A. 1988. Variability of the fluvial system in space and time. In
T. Rosswall, R. G. Woodmansee, and P. G. Risser (eds), *Scales and
Global Change: Spatial and Temporal Variability in Biospheric and
Geospheric Processes*. Chichester: Wiley, pp. 225–50.

Schumm, S. A. 1991. *To Interpret the Earth: Ten Ways to be Wrong*. New
York: Cambridge University Press.

Schumm, S. A. and Lichty, R. W. 1965. Time, space, and causality in
geomorphology. *American Journal of Science* 273: 110–19.

Seiler, F. A. 1986. Use of fractals to estimate environmental dilution factors
in river basins. *Risk Analysis* 6: 15–25.

Seyfried, M. S. and Wilcox, B. P. 1995. Scale and the nature of spatial
variability: field examples having implications for hydrologic modeling.
Water Resources Research 31: 173–84.

Shaw, H. R. 1994. *Craters, Cosmos, and Chronicles: a New Theory of
Earth*. Palo Alto, CA: Stanford University Press.

Simon, A. 1992. Energy, time and channel evolution in catastrophically
disturbed fluvial systems. In Phillips and Renwick, 1992, pp. 345–72.

Simon, A. and Thorne, C. R. 1996. Channel adjustment of an unstable
coarse-grained stream: opposing trends of boundary and critical shear
stress, and the applicability of extremal hypotheses. *Earth Surface Processes and Landforms* 21: 155–80.

Simonson, R. W. 1959. Outline of a generalized theory of soil genesis. *Soil
Science Society of America Proceedings* 23: 152–6.

Slingerland, R. 1981. Qualitative stability analysis of geologic systems with
an example from river hydraulic geometry. *Geology* 9: 491–3.

Small, T. W., Schaetzl, R. J., and Brixie, J. M. 1990. Redistribution and mixing of soil gravels by tree uprooting. *Professional Geographer* 42: 445–57.

Smeck, N. E., Runge, E. C. A., and MacKintosh, E. E. 1983. Dynamics and genetic modelling of soil systems. In L. P. Wilding, N. E. Smeck and G. F. Hall (eds), *Pedogenesis and Soil Taxonomy*. Amsterdam: Elsevier, pp. 51–81.

Smith, T. R. and Bretherton, F. P. 1972. Stability and the conservation of mass in drainage basin evolution. *Water Resources Research* 8: 1506–29.

Stark, C. P. 1991. An invasion percolation model of drainage network evolution. *Nature* 352: 423–5.

Stephens, C. G. 1947. Functional synthesis in pedogenesis. *Transactions of the Royal Society of Australia* 71: 168–81.

Stephens, E. P. 1956. The uprooting of trees: a forest process. *Soil Science Society of America Proceedings* 20: 113–16.

Stevenson, J. C., Ward, L. G., and Kearney, M. S. 1988. Sediment transport and trapping in marsh systems: implications of tidal flux studies. *Marine Geology* 80: 37–59.

Straub, S. 1996. Self-organization in the rapid flow of granular material: evidence for a major flow mechanism. *Geologische Rundschau* 85: 85–91.

Sun, T. and Meakin, P. 1994. The topography of optimal drainage basins. *Water Resources Research* 30: 2599–610.

Sun, T., Meakin, P., and Jossang, T. 1994. A minimum energy dissipation model for drainage basins that explicitly differentiates between channel networks and hillslopes. *Physica A* 210: 24–47.

Sun, T., Meakin, P., and Jossang, T. 1995. Minimum energy dissipation river networks with fractal boundaries. *Physical Review E* 51: 5353–9.

Sutherland, R. A., van Kessel, C., Farrell, R. E., and Pennock, D. J. 1993. Landscape-scale variations in plant and soil nitrogen-15 natural abundance. *Soil Science Society of America Journal* 57: 169–78.

Takayasu, H. and Inaoka, H. 1992. New type of self-organized criticality in a model of erosion. *Physical Review Letters* 68: 966–9.

Tarboton, D. G., Bras, R. L., and Rodriguez-Iturbe, I. 1988. The fractal nature of river networks. *Water Resources Research* 24: 1317–22.

Tarboton, D. G., Bras, R. L., and Rodriguez-Iturbe, I. 1989. Scaling and elevation in river networks. *Water Resources Research* 25: 2037–51.

Tarboton, D. G., Bras, R. L., and Rodriguez-Iturbe, I. 1992. A physical basis for drainage density. *Geomorphology* 5: 59–76.

Tausch, R. J., Wigand, P. E., and Burkhardt, J. W. 1993. Plant community thresholds, multiple steady states, and multiple successional pathways: legacy of the Quaternary? *Journal of Range Management* 46: 439–47.

Thomas, D. S. G. and Middleton, N. J. 1994. *Desertification: Exploding the Myth*. Chichester: John Wiley.

Thompson, C. H. 1983. Development and weathering of large parabolic dune systems along the subtropical coast of eastern Australia. *Zeitschrift für Geomorphologie* suppl. 45: 205–25.

Thompson, C. H. and Bowman, G. M. 1984. Subaerial denudation and weathering of vegetated coastal dunes in eastern Australia. In B. G. Thom (ed.), *Coastal Geomorphology in Australia*. Sydney: Academic Press, pp. 263–90.

Thorn, C. E. and Welford, M. R. 1994. The equilibrium concept in geomorphology. *Annals of the Association of American Geographers* 84: 666–96.

Thornes, J. B. 1985. The ecology of erosion. *Geography* 70: 222–35.

Thornes, J. B. 1988. Erosional equilibria under grazing. In J. L. Bintliff, D. A. Davidson and E. G. Grant (eds), *Conceptual Issues in Archeology*. Edinburgh: Edinburgh University Press, pp. 193–210.

Thornes, J. B. 1990. The interaction of erosional and vegetational dynamics in land degradation: spatial outcomes. In J. B. Thornes (ed.), *Vegetation and Erosion*. Chichester: John Wiley, pp. 41–53.

Tonkin, P. J. and Basher, L. R. 1990. Soil-stratigraphic techniques in the study of soil and landform evolution across the Southern Alps, New Zealand. In L. D. McFadden and P. L. K. Knuepfer (eds), *Soils and Landscape Evolution*. Proceedings of the 21st Binghamton Symposium in Geomorphology. Amsterdam: Elsevier, pp. 547–75.

Tricart, J. 1988. Le sol dans l'environnement écologique. *Revue de Geomorphologie Dynamique* 27: 113–28.

Troch, P. A., Smith, J. A., Wood, E. F., and de Troch, F. P. 1994. Hydrologic controls of large floods in a small basin: central Appalachian case study. *Journal of Hydrology* 156: 285–309.

Troutman, B. M. and Karlinger, M. R. 1992. Gibbs' distribution on drainage networks. *Water Resources Research* 28: 563–77.

Troutman, B. M. and Karlinger, M. R. 1994. Inference for a generalized Gibbsian distribution on channel networks. *Water Resources Research* 30: 2325–38.

Tsonis, A. A. and Elsner, J. B. 1989. Chaos, strange attractors, and weather. *Bulletin of the American Meteorological Society* 70: 14–23.

Tuan, Y-F. 1968. *The Hydrologic Cycle and the Wisdom of God: a Theme in Geoteleology*. Toronto: University of Toronto Press.

Twidale, C. R. 1991. A model of landscape evolution involving increased and increasing relief amplitude. *Zeitschrift für Geomorphologie* 35: 85–109.

Twidale, C. R. 1993. The research frontier and beyond: granitic terrains. In J. D. Vitek and J. R. Giardino (eds), *Geomorphology: the Research Frontier and Beyond*. Proceedings of the 24th Binghamton Geomorphology Symposium. Amsterdam: Elsevier, pp. 187–223.

Van Deusen, P. 1990. Stand dynamics and red spruce decline. *Canadian Journal of Forest Research* 20: 743–9.

Van Es, M. H., Cassel, D. K., and Daniels, R. B. 1991. Infiltration variability and correlations with surface soil properties for an eroded Hapludult. *Soil Science Society of America Journal* 55: 486–92.

Van Pelt, J., Woldenberg, M. J., and Verwer, R. W. H. 1989. Two generalized topological models of stream network growth. *Journal of Geology* 97: 281–99.

Vasenev, I. I. and Targul'yan, V. O. 1995. A model for the development of sod-podzolic soils by windthrow. *Eurasian Soil Science* 27: 1–16.

Verosub, K. L., Fine, P., Singer, M. J., and TenPas, J. 1993. Pedogenesis and paleoclimate: interpretation of the magnetic susceptibility record of Chinese loess-paleosol sequences. *Geology* 21: 1011–14.

Vitousek, P. M. 1994. Factors controlling ecosystem structure and function. In *Factors of Soil Formation: a Fiftieth Anniversary Retrospective*. Madison, WI: Soil Science Society of America Special Publication 33: 87–97.

Wainwright, J., Parsons, A. J., and Abrahams, A. D. 1995. A simulation study of raindrop erosion in the formation of desert pavements. *Earth Surface Processes and Landforms* 20: 277–91.

Walker, B. H. 1993. Rangeland ecology: understanding and managing change. *Ambio* 22: 80–7.

Westoby, M., Walker, B. H., and Noy-Meir, I. 1989. Opportunistic management of rangelands not at equilibrium. *Journal of Range Management* 42: 266–74.

Wilding, L. P. 1994. Factors of soil formation: contributions to pedology. In *Factors of Soil Formation: a Fiftieth Anniversary Retrospective.* Madison, WI: Soil Science Society of America Special Publication 33: 15–30.

Williams, G. P. 1978. Hydraulic geometry of river cross-sections: theory of minimum variance. *US Geological Survey Professional Paper* 1029: 1–47.

Williams, M. A. J. and Balling, R. C. 1996. *Interactions of Desertification and Climate.* London: Edward Arnold.

Willgoose, G. R. 1994. A statistic for testing the elevation characteristics of landscape simulation models. *Journal of Geophysical Research* 99B: 13,987–96.

Willgoose, G. R., Bras, R. L., and Rodriguez-Iturbe, I. 1991a. A coupled channel network growth and hillslope evolution model. 1: Theory. *Water Resources Research* 27: 1671–84.

Willgoose, G. R., Bras, R. L., and Rodriguez-Iturbe, I. 1991b. A coupled channel network growth and hillslope evolution model. 2: Nondimensionalization and applications. *Water Resources Research* 27: 1685–96.

Willgoose, G. R., Bras, R. L., and Rodriguez-Iturbe, I. 1991c. Results from a new model of river basin evolution. *Earth Surface Processes and Landforms* 16: 237–54.

Wilson, J. B. and Agnew, A. D. Q. 1992. Positive-feedback switches in plant communities. *Advances in Ecological Research* 23: 263–336.

Woldenberg, M. J. 1969. Spatial order in fluvial systems: Horton's laws derived from mixed hexagonal hierarchies of drainage basin areas. *Geological Society of America Bulletin* 80: 97–112.

Wolf, A., Swift, J. B., Swinney, H. L., and Vastano, J. A. 1985. Determining Lyapunov exponents from a time series. *Physica D* 16: 285–317.

Wright, L. D. and Short, A. D. 1984. Morphodynamic variability of surf zones and beaches: a synthesis. *Marine Geology* 56: 93–118.

Wright, P. B. 1980. An approach to modeling climate based on feedback relationships. *Climatic Change* 2: 283–98.

Wyrick, M. 1993. Soil profile development and geomorphic surfaces in the North Carolina coastal plain. MA thesis, East Carolina University, Greenville, N. C.

Xue, Y. and Shukla, J. 1993. The influence of land surface properties on Sahel climate. I: Desertification. *Journal of Climate* 6: 2232–45.

Yaalon, D. H. 1975. Conceptual models in pedogenesis: can the soil-forming functions be solved? *Geoderma* 13: 189–205.

Yair, A. 1990. The role of topography and surface cover upon soil formation along hillslopes in arid climates. In L. D. McFadden and P. L. K. Knuepfer (eds), *Soils and Landscape Evolution.* Proceedings of the 21st Binghamton Symposium in Geomorphology. Amsterdam: Elsevier, pp. 287–300.

Yair, A. 1992. Climate change and environment at the desert fringe, northern Negev, Israel. In K-H. Schmidt and J. De Plocy (eds), *Functional Geomorphology. Catena* suppl. 23: 47–58.

Yang, C. T. 1992. Force, energy, entropy, and energy dissipation rate. In V. P. Singh and M. Fiorentino (eds) *Entropy and Energy Dissipation in Water Resources.* Dordrecht: Kluwer, pp. 63–89.

Zdenkovic, M. L. and Scheidegger, A. E. 1989. Entropy of landscapes. *Zeitschrift für Geomorphologie* 33: 361–71.

Zeng, X., Pielke, R. A., and Eykholt, R. 1993. Chaos theory and its applications to the atmosphere. *Bulletin of the American Meteorological Society* 74: 631–44.

Zinke, P. J. 1962. The pattern of individual forest trees on soil properties. *Ecology* 43: 130–3.

Index